IMAGINE, if you will, a banking system unlike any other. Every morning, precisely at midnight, your account is credited with exactly twenty-four hours. You cannot draw against tomorrow. You cannot save for the weekend. There is no overdraft protection. This is the *daily miracle* of time.

Yet, if you are like many modern practitioners, you wake up each morning feeling a heaviness, a flicker of a nagging dread. You sense the presence and the stealth of a talented pickpocket that covets your time and will rob you of it without a second of regret.

An influencer will tickle your guilt and offer tools and apps to cure all your lack of time ills. But it is not a time management problem. You are the victim of a systemic robbery. *The Great Time Delusion* exposes the lie at the heart of the digital age. That lie is the promise that technology will save us time. If you are exhausted by the relentless demands of an always on world, secure your copy of this book. Take the first step toward reclaiming your most precious of all your assets, time, and the rewards that come with it.

Consider the events of July 19, 2024. On that day, a single, faulty software update from CrowdStrike rippled across the globe. Reports of grounded flights, paralyzed hospitals, and frozen financial markets entered the 24 hour news loops. In an instant, the daily miracle was suspended for millions. This event perfectly illustrates the invisible crisis of *technical friction* and the *update trap*. We have traded our autonomy for imaginary convenience, and in doing so, we have amassed a burdensome *time debt*. We no longer own our days; we lease them from the attention economy, digital platforms and willfully manipulative algorithms.

More than a century ago, in 1910, Arnold Bennett warned of the dangers of letting time slip away without intention. His Edwardian wisdom was stunningly profound, but he could never have anticipated the weapons of mass distraction we face today. *The Great Time Delusion* bridges this gap. It takes the foundational philosophy of living deliberately and fortifies it with modern, actionable strategies to combat the time thieves of the 21st century.

This is not another generic productivity manual filled with hacks to help you squeeze three more emails into your lunch break. It is a declaration of independence. Within these pages, you will discover the *three pillars of protection*. You will learn how to identify the *threshold problem* that keeps you perpetually distracted and how to navigate the very real, physiological weight of telepressure.

Reclaiming your 24 hours *is not* about doing more. It is about owning more of *your* time. It is about deciding that you are the Chief Executive Officer of your own soul. You possess exactly the same number of hours in a day as Albert Einstein, Benjamin Franklin, and Earnest Hemingway. The difference lies entirely in how you protect those hours from the systems designed to steal hours and days and months from you.

The daily miracle is yours to claim, but the defense of it must be engineered by you. Enjoy *The Great Time Delusion* today and start building the life you actually want to live, one fiercely protected hour at a time.

For permission requests, contact the author at:
[dave@armerclad.com]

Disclaimer:

This is a work of non-fiction. The author has made every effort to ensure the accuracy of the information within this book was correct at time of publication. The author does not assume and hereby disclaims any liability to any party for any loss, damage, or disruption caused by errors or omissions, whether such errors or omissions result from accident, negligence, or any other cause.

ISBN - 978-1-7350164-5-0
Printed in the United States of America

First Edition: May 2026

Disclosure: This manuscript was developed with the assistance of AI-based tools used solely for preliminary tasks such as brainstorming, outlining, drafting, and grammar support. All content has been extensively reviewed and/or rewritten and refined by the author to ensure clarity, accuracy, and fidelity to the intended message. Every factual statement, citation, and reference has been independently verified through traditional research methods and validated using professional tools such as Edifix. The final work reflects the author's judgment, interpretation, and editorial decisions, with AI serving only as a supplementary aid rather than an infallible source of authoritative information.

Introduction

The Man from the Potteries

ENOCH ARNOLD BENNETT (1867–1931) was not born into the leisure class. He was forged in the smoke and clay of the Potteries in Staffordshire, England, a grim industrial district where time was measured in factory whistles and shift changes. He began his professional life not as a literary lion, but as a solicitor's clerk in London, earning a pittance and living the very life of "eight hours for the desk, eight hours for sleep" that he would later seek to reform.

Wiki-Creative Commons

Through sheer force of will, and a discipline that bordered on the ascetic, Bennett transformed himself into one of the most prolific and celebrated men of letters of the early 20th century. He wrote over thirty novels, including the undeniable masterpiece *The Old Wives' Tale* (1908), which established him as a giant of realism alongside Wells and Conrad. He was a man of immense public stature, a literary celebrity who dined with cabinet ministers and shaped public opinion. Yet, true to his fiercely independent spirit, he famously refused a knighthood in 1918, preferring the sovereignty of his own name to a title bestowed by the crown.

But it was a slender volume published in 1910 that would become his most enduring legacy. *How to Live on 24 Hours a Day* was not a novel, nor was it a piece of high criticism. It was, in his words, a piece of "pocket philosophy". It was a plea to the practitioner to realize that the sixteen hours outside of their employment were not merely residual time, but the very essence of life itself. Bennett saw, even then, that men and women were sleepwalking through their existence, waiting for a retirement that might never come.

Bennett's Blueprint Revisited

Bennett wrote for a world of telegrams, tramcars, and evening newspapers. He could not have foreseen the internet, the smartphone, the algorithm or artificial intelligence. Yet his diagnosis of the human condition remains *terrifyingly* accurate. He identified the centrifugal force that pulls our attention apart. He identified the "prig" who uses self-importance as a weapon. He identified the tragedy of the unexamined day.

Bennett was the inspiration for this book. They say hindsight is 20/20. Having survived into my seventh decade my hindsight may be 20/15 by now. This manuscript is not a rewrite of Bennett's work; it is a vision, a resurrection of his logic. A background in labor relations has made me painfully aware of the struggle of the blue collar and professional classes (henceforth referred to as practitioners) as they cope with leadership that values authority over process sanity. I have taken the essence of Bennett's 1910 arguments and used it to build a refuge for the digital age. Where Bennett spoke of the evening paper, we speak of the infinite scroll. Where he spoke of mental concentration we speak of being in the zone. The antagonist has changed its face, it is faster, louder, and more intentionally addictive, but the battleground remains the same; the human mind, human time. I believe Bennett deserves the mantle of prophet.

This book is also the result of a partnership. It is a collaboration between myself, author and editor, and the machine that is driving as many dilemmas as solutions. I am Hamilton Armer, the architect of this project. I leaned on an amazing research team called Gemini and ChatGPT.

Gemini, ChatGPT and their powerful Large Language Models (LLM) have utilized processing power to scan the century of research that has passed since Bennett's death. It has pulled from neuroscience, psychology, and modern economics (among other fields) to validate Bennett's intuitions. Together we have synthesized the wisdom of Stoics, the data of productivity experts, and the warnings of digital minimalists. The references display the breadth and depth of that knowledge. I feel that I am standing on the shoulders of giants.

A note of AI caution however. AI deals in probabilities. It has no real awareness of the binary choices of true or false like the code that runs our devices. When faced with multiple choices it will take the most probable, not the most truthful. AI will lie to your face without even the possibility of guilt or remorse. I learned the hard way that if it can't find a piece of research you desire, it will just make one up to please you. I will offer a shameless plug to Edifix.com for vetting AIs lies. Approach AI in the same fashion you should approach a politician. A 500-word prompt that begs for 100% accuracy will not keep AI honest. I do find it an incredible tool for brainstorming, offering options and breaking writers block. It is like having a brainstorming team and a research team at your beck and call. AI led me to Arnold Bennett. His work inspired me and this book. Just remember, AI has neither conscience nor scruples. It will not value your time or regret any consequences. Use it in your own pursuits with a large helping of caveat emptor.

AI aside, my role has been to function as the bridge between Bennett's analog wisdom and 21st century digital reality. We have created an updated blueprint of his chapters, not to replace the original human voice, but to amplify a voice that has been drowned out by the noise of the digital labyrinth.

I honor Arnold Bennett not by treating his work as a museum piece, but by testing it, forging it and tempering it. I honor him by hypothesizing that even in an age of artificial intelligence, **the human mind remains the most important thing worth optimizing.**

The 24 hours per day are yours. I hope this will empower some portion of humanity to compete in the game of life more effectively (as opposed to more efficiently). I encourage you to peruse the essays within the addendum as they shine a light on the contemporaneous literary tensions of Bennett's era. I was unamused to learn that Virginia Woolf despised Bennett's work, looking down her nose from her financially privileged perch. Bennett did it the old fashioned way; he earned it by the sweat of his brow.

Essential Prologue

The Architecture of Invisible Labor

THE 24 hour day is the only truly democratic resource left on earth. It is an unyielding constant that connects the Edwardian clerk of 1910 to the harassed practitioner of the algorithmic 21st century. Yet, despite the undeniable rigidity of this temporal container, the modern experience of time is defined by a paradox. We possess tools of unprecedented speed and efficiency, but we find ourselves with less discretionary time than the ancestors who labored without them. The explanation for this phenomenon does not lie in the physics of time (although time is relative) but in the economics of attention. Instead of sufficient time spent creating things, we spend time monitoring, editing, and managing a massive amount of low value information. It is a crisis rooted in the invisible accumulation of unfulfilling labor, a concept that has mutated from a domestic necessity into a digital pathology.

To understand why the modern day feels so compressed, one must trace the lineage of the activities that fill it. The concept of unpaid labor that supports a formal economy was originally described by Ivan Illich. He defined *shadow work* as the unpaid and unacknowledged complement to industrial production (Illich, 1981). This category historically included activities such as commuting to a workplace, shopping for provisions, or performing domestic chores. Illich argued that these tasks were a structural requirement of modern society because they allowed the wage earner to participate in the formal workforce. The factory could not function unless the worker arrived fed, clothed, and rested, yet the labor required to produce that readiness never appeared on a corporate spreadsheet or quarterly report. While these activities are necessary for the economy to function, they are not compensated, and they often prevent individuals from pursuing personal growth.

For the industrial worker, shadow work was physical and geographic. It was the miles walked to the factory or the coal scrubbed from a uniform. However, as the economy shifted from heavy industry to information, the nature of this unpaid burden evolved. Craig Lambert updated this concept for the digital era by examining how technology reallocates labor. He defined shadow work as the unpaid tasks that corporations transfer to consumers through various technological interfaces (Lambert, 2015). This process frequently removes professional middlemen and requires individuals to handle responsibilities like self-checkout at retail stores, the tagging of their own baggage at the airport, or the management of complex online banking profiles.

Lambert observed that although these digital tools are marketed as conveniences that offer autonomy, they often increase the total volume of administrative labor that an individual must perform. We are no longer merely customers. We have become the unpaid data entry clerks, travel agents, and logistics managers for the very companies to which we pay money. This manuscript builds on these historical perspectives to explain how modern technical friction reduces the time available for meaningful personal output. In this context, the term refers to the hidden administrative tasks and digital maintenance that hinders a person from engaging in high value, direct personal production. This type of invisible labor typically involves managing constant software updates, troubleshooting connectivity issues, or responding to the relentless ping of digital notifications and the algorithmic drumbeat of the *attention economy*.

This evolution brings us to a critical juncture that Arnold Bennett could not have foreseen. In 1910, Bennett wrote *How to Live on Twenty Four Hours a Day* with a singular, optimistic goal. He sought to awaken the worker from an "uneasy doze", urging them to carve out seven and a half hours from their week for poetry, history, and the cultivation of the spirit (Bennett, 1910). Bennett believed that if the humans could simply manage their fatigue and organize their minds, they could claim a stature equal to that of the aristocracy. He assumed that the enemies of this pursuit were internal sloth and the natural limits of human energy.

In the 21st century, the enemies of the sovereign mind are external, highly capitalized, and both politically and economically motivated. The goals Bennett reached for, intellectual autonomy, reflective silence, and the accumulation of knowledge, have been weaponized by the attention economy. The time Bennett begged us to protect has been colonized by a new form of shadow work that is far more insidious than the physical chores Illich described.

The modern political and economic apparatus has discovered that a fragmented mind is a profitable mind. By flooding the cognitive space of our minds with headline stress and algorithmic outrage, these systems can force the individual into a state of permanent reactivity. The citizen is no longer reading history to understand the present, as Bennett advised. Instead, they are doomscrolling through a curated feed of crises or 24 hour news cycles. We are performing the unpaid emotional labor of anxiety that keeps the platform's engagement metrics high. This is not leisure, and it certainly is not work. It is a grey zone of cognitive extraction where the user's attention is harvested to sell advertisements or manufacture political consent or dissent (Wu, 2016).

This exploitation creates a devastating feedback loop. The individual seeks to use technology to become more efficient, hoping to reclaim the time futurists promised. Yet the tools of efficiency, the smartphone, the inbox, the news alert, bring with them a taxation of shadow work that exceeds the time they save, sometimes by multiples. The user spends hours managing the digital environment, curating their professional avatar, their persona, and responding to the artificial urgency of the network. The result is a population that is technically connected but intellectually insolvent.

By understanding the shift from physical maintenance to digital task loading, individuals can begin to identify the time debt that restricts their intellectual sovereignty, intellectual development and subsequentially, their most precious resource, time. We have moved from the physical shadow work of the commute to the cognitive shadow work of the interface. The tragedy is that while the industrial worker knew they were tired because they had walked home, the digital practitioner is fatigued without having moved an inch. They have spent their day spinning the wheels of an attention economy that views their exhaustion not as a bug, but as a feature. This manuscript serves as a field manual for dismantling that architecture, rejecting the false promise of digital convenience, and reclaiming the 24 hours that are, by rights, the only territory we truly own.

Chapter One

The 24 Hour Bank: The Daily Miracle

IMAGINE a banking system unlike any other. Every morning, precisely at midnight, your account is credited with exactly 24 hours. There is no overdraft protection. There is no rollover of unused minutes. If you do not spend your balance, it is permanently and irrevocably deleted at the stroke of the next midnight.

This bank account is the only truly impartial resource left in our global society. In a world defined by the digital divide (online vs. offline populations) and varying access to hardware and capital, time remains the ultimate equalizer. It cannot be stolen by the unscrupulous, it cannot be taxed by a central authority, and it cannot be upgraded by a software patch. It is yours to spend, or to let evaporate. Unfortunately, we treat this bank account with a carelessness that would be unthinkable with our financial assets. We allow micro withdrawals to drain our balance throughout the waking day. We let a notification ping steal five minutes here; we let a social media algorithm steal twenty minutes (or more) there. By the time we reach the end of our waking hours, we look at our empty vault and ponder where the fortune went. As Bennett (1910) famously noted, your purse is magically filled with 24 hours of the "unmanufactured tissue of the universe", yet we spend it as if the supply were infinite.

Each of us has been granted a daily miracle; the immutable credit of 1,440 minutes received every 24 hours. It can be described as the only absolute, unequivocally equitable resource, unmanufacturable and unstealable (absent unjust incarceration).

We are currently living through the most profound era of the Great Time Delusion; the societal hallucination that new technologies will grant humans more leisure, when it historically only increases the tempo and the pace of existence. The delusion is that the next software update, the next automated workflow, or the next leap in artificial intelligence will finally grant us the additional leisure or personally productive time we crave. We have been sold a promise of efficiency that has, in reality, served to increase our frustration and further enrich the global plutocracy. As technical systems accelerate, human experience has not expanded, it has been compressed. We are moving faster than ever, yet many of us feel more behind than any generation in recorded history. Welcome to the concept of technical friction. Technical friction is the significant drain on time caused by the perpetual flux of digital environments, such as algorithms, critical patches, updates, notifications, news feeds or relearning optimized interfaces.

The Update Trap, The CrowdStrike Outage

A visceral example of technical friction occurred on July 19, 2024, when a single, faulty software update from the cybersecurity firm CrowdStrike reportedly disabled approximately 8.5 million Windows devices worldwide (Microsoft, 2024). This event captured the global eye as it effectively froze the daily miracle for millions. Instead of automated security promising peace of mind, it created a massive time debt where practitioners across airlines, hospitals, and financial institutions spent days manually troubleshooting systems.

What is a time debt? The concept is derived from Cunningham's (1992) metaphor of *technical debt*, which suggests that opting for immediate ease over long-term maintenance creates a cumulative deficit that must eventually be paid back with interest. In this case, the maintenance of the tools became the primary task itself.

The Southwest Airlines Holiday Meltdown (2022)

The Debt - Deferred Maintenance, Archaic Systems: For decades, Southwest Airlines operated on a lean point-to-point model while allegedly neglecting to upgrade its internal crew scheduling software, known as SkySolver (Helmrich et al., 2023). The leadership purportedly prioritized customer facing improvements and short term operational efficiency over the invisible work of modernizing backend infrastructure (Tufekci, 2022). This was a classic accumulation of time debt; they borrowed time by skipping complex upgrades to keep planes moving in the short term.

The Consequence: In December 2022, a severe winter storm triggered a system collapse that SkySolver apparently could not handle. Because the software required manual data entry during disruptions, the airline lost track of its pilots and flight attendants (Helmrich et al., 2023). The result was over 16,900 canceled flights and a financial impact exceeding $1 billion, including a $140 million federal penalty (U.S. Department of Transportation, 2023). The time they saved by not upgrading the system was lost exponentially in a single week of technical friction.

The Efficiency Paradox

Throughout 2025, the widespread integration of generative AI agents led to what economists call the Jevons Paradox of productivity. The Jevons Paradox originated in the 19th century during the Industrial Revolution. It was first identified by economist William Stanley Jevons, who observed that as steam engines became more efficient at burning coal, the total national consumption of coal actually increased rather than decreased. The paradox posits that technological progress that increases the efficiency of a resource does not lead to a decrease in its total consumption; instead, by lowering the relative cost of using that resource, it drives an exponential rise in demand and total consumption (Jevons, 1865).

The LED Rebound Effect

The Expectation, Efficiency = Savings: In the early 2010s, a global push began to replace inefficient incandescent bulbs with solid state lighting. The logic was simple; because LEDs are up to 85% more efficient than traditional bulbs, their widespread adoption was predicted to drastically reduce global energy consumption for lighting (Tsao et al., 2010).

The Reality, The Rebound Effect: Historical data analyzed by Tsao et al. (2010) revealed a consistent economic truth. As the cost of light (both in energy and dollars) drops, humanity does not bank the savings. We simply consume more light. Because LEDs made lighting so cheap, society moved the goalposts. We began illuminating areas that were previously left dark, building exteriors, empty parking lots at 3:00 AM, decorative landscapes, and installing higher concentrations of lumens in our homes (Tsao et al., 2010). Consequently, while the efficiency of the technology improved, the total global consumption of energy for lighting continued to rise rather than fall, a textbook manifestation of the Jevons Paradox.

The Paperless Office Paradox

The Expectation, Digital Substitution: With the advent of the personal computer and email in the 1980s and 90s, futurists predicted the imminent death of paper. The logic was, because digital documents are weightless, searchable, and instantly transferable, the high cost of physical printing would naturally drive paper consumption to near zero (Sellen & Harper, 2002).

The Reality, The Ease of Generation: Instead of replacing paper, computers simply lowered the transaction cost of creating it. Before the PC, typing a report required a typewriter and significant manual effort. Corrections were punishing. The computer made drafting, editing, and printing far less tedious. Consequently, office workers began printing out multiple drafts for minor edits, reading copies for better retention, and printing emails for archiving (Sellen & Harper, 2002).

Global consumption of office paper did not vanish, it actually rose. The digital tool did not eliminate the physical artifact; it merely acted as a hyper efficient printing press. It proved Jevons' rule that efficiency increases consumption. This creates a concept called *frictionless waste.* When a task becomes frictionless (like hitting print or CC All), we stop treating the resource (paper or attention) as scarce. If it is now free to invite 10 people to a Zoom meeting, we invite them all, creating hundreds of hours of collective time debt without a second thought.

The $4.6 Billion Data Entry Crisis

The Shift, From Scribes to Screens: In the last decade, healthcare systems transitioned to Electronic Health Records (EHR) to improve efficiency. In doing so, they reduced the roles of the medical transcriptionist or scribe. The task of documenting care, typing notes, entering billing codes, and managing checkboxes, was often offloaded onto the doctor (Sinsky et al., 2016).

The Shadow Work Load: Research by the American Medical Association found that for every one hour a physician spends with a patient, they now spend two hours on EHR and desk work (Sinsky et al., 2016). This is the definition of shadow work, the invisible administrative labor required to support the real job, which has now eclipsed the job itself.

The Financial Consequence: This is not just an annoyance; it is a catastrophe. The accumulation of this shadow work is the primary driver of physician burnout. A landmark study estimated that this burnout costs the U.S. healthcare system approximately $4.6 billion annually in physician turnover and reduced clinical hours (Han et al., 2019). By saving money on administrative support, the system incurred a multibillion dollar debt in lost expertise and efficiency. Ironically, wait times for appointments and services have steadily and dramatically increased as proponents of market driven health care for profit hold sway (AMN Healthcare, 2025).

A fundamental crisis of the 21st century is the growing gap between our slowly evolving biology and our exponential technology growth. While our tools have become capable of processing trillions of operations per second, the human prefrontal cortex remains tethered *to the same pace it occupied centuries ago*. Every person, from the lead architect of a global cloud infrastructure to a cubicle end user struggling with a software update, is bound by the same immutable physics of the clock. As Bennett (1910) argued, "we have, and we have always had, all the time there is".

Realize that we need more than a productivity strategy; we need a radical act of intellectual sovereignty. If you find yourself saying, "I'll focus on my personal needs once the current development cycle ends", you are living in a state of self-imposed cognitive bankruptcy. The cycles will *never* end. The updates will *never* stop. If you do not manage your time today, you are effectively forfeiting your consciousness to the momentum of a machine that never sleeps (Rosa, 2013).

The Update Trap, Tools Become Tasks

In the 21st century, the most significant time thief is often the very technology marketed as a timesaver. Consider the phenomenon of the software update, a perfect metaphor for the friction of contemporary existence. In an era of continuous data integration and continuous software deployment, our environment is in a state of perpetual flux.

We have all experienced the update trap. You sit down to engage in deep, meaningful work, but your operating system demands a restart. A critical security patch is required. Suddenly, the interface you have mastered has been optimized. Buttons have moved. Features you relied on have been renamed or hidden behind new menus. You spend the next hour relearning how to use your own tools. This is technical friction, and it represents a significant drain on our 24 hour bank.

The cost can be stunning. Revisiting the CrowdStrike update drives home the point. CrowdStrike had released a routine content configuration update, specifically "Channel File 291", to its Falcon sensor software. The update was intended to help systems target malicious named pipes (used by hackers to gain access)(CrowdStrike, 2024). However, the update contained a logic error that the validation software failed to catch. Because modern IT systems are designed for frictionless updates, this faulty file was automatically pushed to millions of Windows machines simultaneously, bypassing the traditional "canary" testing phases that would have isolated the error (Microsoft, 2024).

Within minutes, millions of Windows devices entered a boot loop, displaying the infamous Blue Screen of Death (Microsoft, 2024). The impact was not just digital but physical. Flights were grounded, hospitals cancelled non-urgent surgeries, and 911 services went offline.

The financial toll was unprecedented for a single software update. Delta Airlines alone reported a $500 million impact. That figure was confirmed by CEO Ed Bastian following the outage (Josephs, 2024). Insurers estimated the total direct financial loss to Fortune 500 companies at roughly $5.4 billion (Insurance Journal, 2024). The update trap here was the assumption that speed of updating equals security. In reality, the speed of the update mechanism was the primary cause of the destruction.

We are constantly updating our environment but far less often upgrading our lives. This friction creates a state of cognitive switching cost, where the mind is forced to recalibrate every time the digital landscape shifts (Carr, 2010). We spend too much of our daily miracle troubleshooting our tools rather than using them to inhabit our lives. We are so tied up maintaining the machinery of our existence that we have forgotten how to drive the vehicle toward a meaningful destination.

The Second Industrial Revolution of the Soul

To understand why we struggle with this 24 hour bank, we must look at the system's history of time. Before the mid-19th century, time was a local and fluid concept. Every town set its own clock based on the position of the sun. It was the Transcontinental Railroad that forced the standardization of time so that trains would not collide on shared tracks.

The Kipton Disaster (April 18, 1891) On a spring morning in Kipton, Ohio, two trains, the Lake Shore and Michigan Southern Railway's prestigious Fast Mail and an accommodation train (local passenger train), collided head on at full speed. The investigation revealed a chillingly simple cause. The conductor of the accommodation train had a watch that had stopped for exactly four minutes before starting again. He was unaware of the discrepancy. Believing he had plenty of time to reach a siding before the Fast Mail arrived, he stayed on the main track. The resulting impact killed eight people, including both engineers and several postal clerks (Bartky, 2000).

The Legacy, The Railroad Approved Standard: This horror story became the catalyst for the Ball Railroad Time Inspection System. The industry realized that having Standard Time Zones was useless if the mechanical devices measuring that time were unreliable. Following this disaster, Webb C. Ball was appointed to establish strict standards for railroad approved watches, requiring them to be accurate to within 30 seconds per week, effectively creating the first high precision technical update in the history of human timekeeping (Bartky, 2000).

We didn't just synchronize our watches; we synchronized our biological rhythms to a mechanical schedule (Gleick, 1999).

Today, we are navigating a Second Industrial (Technical) Revolution. But the factory is no longer a physical building. It is the device in your pocket. The standardized time of the railroad has been replaced by the real time of the internet. Our 24 hours are no longer divided by a factory whistle. They are fragmented by push notifications. The most well known and arguably most chilling example of a push notification in the 21st century is the 2018 Hawaii False Missile Alert.

On the morning of January 13, 2018, at 8:07 a.m. local time, residents and tourists across the Hawaiian Islands received a high priority emergency alert on their smartphones. The message, delivered via the wireless emergency alert system, appeared as a standard system level push notification that bypassed the "Do Not Disturb" settings. It read: **"BALLISTIC MISSILE THREAT INBOUND TO HAWAII. SEEK IMMEDIATE SHELTER. THIS IS NOT A DRILL."**

For 38 minutes an entire state lived in a state of existential terror. People said goodbye to loved ones, hid in bathtubs, and climbed into storm drains. The notification was later revealed to be the result of a technical error. An employee at the Hawaii Emergency Management Agency had mistakenly selected the live alert option from a dropdown menu during a routine internal drill instead of the test option (Federal Communications Commission [FCC], 2018).

The 4:45 AM Florida Wake Up Call (2023), The Disaster: On April 20, 2023, millions of Florida residents were jolted awake at 4:45 AM local time by a piercing emergency alert on their phones. It was not a hurricane or a nuclear missile. It was a test.

The Cause: The Florida Division of Emergency Management purportedly had contracted a third party vendor, Everbridge, to handle the alerts. The vendor supposedly sent the wrong technical specifications, treating a routine monthly test (usually reserved for TV broadcast) as a live emergency alert to be pushed to wireless devices (Mazzei, 2023).

The Consequence: The frictionless nature of the system meant that a single unchecked box bypassed all social norms of sleep and privacy. The backlash was so severe that Governor Ron DeSantis ordered an immediate investigation, and the state severed its contract with the vendor shortly thereafter (Mazzei, 2023). It proved that when you remove the friction of manual review (two persons preferably), you also remove the safety valve of common sense.

The Samsung 1 Worldwide Glitch (2020), The Disaster: In February 2020, millions of Samsung Galaxy users worldwide received a push notification from the "Find My Mobile" app. The notification contained no text, only the number 1 stacked on top of another number 1.

The Cause: This was not a hack, but an internal test. A developer had unintentionally triggered a test notification to the live database rather than the internal test group (Kastrenakes, 2020). Because the system was designed for global scale, the error propagated instantly to devices from Tokyo to Toronto.

The Consequence: While physically harmless, it triggered a massive wave of panic. Users believed their phones had been hacked or compromised. Thousands of hours were lost (time debt) as users frantically changed passwords and called support lines (Kastrenakes, 2020).

The Seoul Air Raid False Alarm (2023), The Disaster: On May 31, 2023, at 6:41 AM, the Seoul metropolitan government sent a presidential alert to all mobile phones in the capital: "Prepare to evacuate. Allow children and the elderly to evacuate first."

The Cause: North Korea had launched a space satellite that failed and fell into the sea. The alert system in Seoul, reacting to the launch, issued a blanket wartime evacuation order without specifying why or where to go (Choe, 2023).

The Consequence: The vagueness of the message caused chaos. Naver (Korea's Google) crashed due to traffic as millions tried to find out what was happening. Twenty minutes later, the Ministry of the Interior had to send a second correction alert stating the first was a mistake. That left millions of citizens with alert fatigue, a dangerous state where future legitimate warnings might be ignored (Choe, 2023).

These events serve as the ultimate modern parallels to the 1891 Kipton Disaster. Just as a four minute watch discrepancy caused a physical train wreck in the 19th century, a single user interface error in the 21st century caused a psychological wreck and massive time debts for millions of people. It highlights the terrifying power of the push architecture, where a system can forcefully withdraw peace of mind from the user's 24 hour bank without warning or consent (FCC, 2018).

We have become machines of high efficiency during our practitioner hours, navigating intricate technical architectures and rapid fire deployments. But the moment we clock out, we regress. We have mastered the art of managing technical debt in our practitioner lives, but we have become illiterate in managing it in our own life equity. We are high performing cogs who do not know how to effectively progress once the machine stops turning.

The Sixteen Hour Audit

The modern practitioner tends to view the workday, the eight hours (if you're lucky) sold to an employer, as the day. Everything else is treated as postscript or recovery. This is a catastrophic failure of *life architecture.*

If you sleep for eight hours and work for eight hours, you are left with eight hours of unallocated time equity. That is 2,920 hours a year. In the 21st century, those 2,920 hours are the primary battleground of your life. If you do not claim them, they will be colonized by the attention economy. The algorithm does not want you to learn a new skill, study history, or master a complex system. It wants you to remain in a state of passive consumption, a digital waiting room where your time is the product being sold to advertisers or worse.

For instance, doomscrolling is the persistent, habitual consumption of negative news through scrolling endless, algorithmically sorted feeds. It is the ultimate update trap of the soul, where the user believes they are seeking information to gain control over a crisis, but instead, they are surrendering their cognitive equity to a loop of anxiety. Cable news plays the same game.

The Negativity Bias: Humans are inherently more responsive to negative stimuli as a survival mechanism. Modern algorithms leverage this negativity bias to optimize engagement time. It prioritizes emotionally intense or alarming headlines because they are mathematically more likely to capture a user's attention (Baumeister et al., 2001).

The Infinite Scroll: Unlike the accommodation train of the 19th century, which eventually reached a station, the digital feed has no stopping cue. This design feature eliminates the natural friction that would normally prompt a user to audit their 24 hour bank, leading to lost time unawareness (Kaya & Griffiths, 2025).

The Crisis of 2020: The phenomenon exploded during the COVID 19 pandemic, where the real time updates of the virus's spread created a feedback loop of hypervigilance. Users spent hours in a state of passive consumption, effectively depositing their peace of mind into a vault they could no longer access. This practice is not harmless idleness. Research has linked doomscrolling to *headline stress disorder*, emotional exhaustion, and cognitive overload. It is a state where the user becomes a passive subject of their own technology, watching their 1,440 minutes evaporate into a curated stream of algorithmic digital dread masquerading as awareness.

We are living in a state of permanent vigilance that psychologist Steven Stosny named headline stress disorder, a cyclical loop of anxiety triggered not by immediate physical danger, but by the relentless intrusion of global crises into our private sphere (Stosny, 2017). Unlike natural stress, which resolves when the tiger leaves the cave, this digital stress is sustained by the glowing rectangle in our pocket (or on the TV stand), keeping our cortisol levels elevated even while we sleep.

The Threshold and the Blur

The Threshold Problem: For those who work onsite, the transition from work to home is often filled with cognitive residue, the mental replay of meetings and read or unread messages.

The Blur Problem: For those in hybrid or remote roles, the walls have vanished entirely. The 24 hour bank becomes a single, gray marathon of semi-work. Without a strict regulatory framework for your private hours, the update *never* finishes, and the work *never* truly ends (Newport, 2016).

These issues clearly have real life consequences:

The Residue Fatality, The 2017 USS John S. McCain Collision: The collision between the USS John S. McCain and a commercial tanker is a noteworthy example of what happens when technical friction meets cognitive residue. The crew was struggling with a complex, newly installed touchscreen steering system. Because the interface was unintuitive, it created attention residue. The sailors were so focused on troubleshooting the update of their controls that they lost situational awareness of the physical world. This failure resulted in the deaths of ten sailors. The subsequent investigation highlighted that the complexity of the user interface was a primary factor in the crew's inability to manage their operational window (National Transportation Safety Board, 2019).

The Update Crash, The Knight Capital Group Bankruptcy: In the world of high frequency trading, time is measured in microseconds. In 2012, Knight Capital Group experienced the ultimate update trap. A technician failed to deploy a new software update to one of eight servers, leaving old zombie code active. When the markets opened, the mismatched systems entered a feedback loop executing millions of unintended trades. In just 45 minutes, the firm lost $440 million. It bankrupted a company that had taken years to build. This remains the definitive case study on how a failure to manage the technical update can lead to the total evaporation of a multimillion dollar enterprise in less than a single hour (Securities and Exchange Commission, 2013).

At the heart of both disasters lies a fundamental contradiction. The tools designed to reduce human error often create new, more complex entry points for systemic instability and new errors. The USS John S. McCain represents the failure of *interface sovereignty*. Interface sovereignty is the right and disciplined practice of configuring digital tools so that the user initiates every interaction, rather than the tool initiating it. For the sailors, the move from physical, tactile controls to a complex digital touchscreen was marketed as an upgrade. However, it introduced a layer of technical friction that paralyzed the crew's intuition. Philosophically, this suggests that the more we mediate our reality through screens, the more we risk losing our situational awareness (texting while driving?). That situational awareness is the biological connection to the 24 hour bank we are supposed to be managing. The best practice for an interface (phone, PC, TV) is: Do not look at the screen until you have decided what you are looking for. This can only result from exercising and sharpening the *concentration core*. More on the concentration core to come.

Knight Capital Group serves as a warning about the infinite speed of errors. In the manual era, a mistake was bound by the speed of a human hand. In the era of the update, an error can bankrupt a legacy in less time than it takes to finish a cup of coffee. It proves that our daily miracle is now vulnerable to automated depletion, where a single oversight in a technical patch can liquidate our entire account before we even realize the vault is open.

Mind Control, The Concentration Core

The ultimate tool for reclaiming your miracle is concentration. Bennett (1910) argued that the mind is a muscle that must be disciplined through deliberate exercise. In an era of infinite scrolls, 24 hour news cycles, and bite sized content, the ability to focus on a single, complex topic for ninety minutes is a literal superpower.

When you analyze a complex system, whether it is the historical engineering of a camera lens or the evolution of software architecture, you are not just learning facts. You are building the concentration core necessary to resist the algorithm. A disciplined mind sees the extraordinary in the humdrum. It finds the philosophical depth in a technical manual and the historical wonder in a simple software update. Attention is the alchemy that turns the mundane into the miraculous. Where the distracted mind sees only administrative friction, the disciplined mind sees an artifact of human ingenuity. That mind finds the deep structure of logic in a user guide and the rhythm of progress in a version number. It recognizes that while the tools of the 21st century change every six months, the rules of the 24 hour bank have remained unchanged for millennia.

The goal of this *mind control* is to transition from being a user of time to being a manager of it. As Bennett (1910) argued, there is no greater social irony than the individual who dedicates merely a third of their day to making a living, yet insists they have no time left to actually live.

The Call to Management

Reclaiming your 24 hours is not about doing more. It is about owning more. It is about deciding that you are the CEO of your own soul.

The technical world will continue to change at a dizzying pace. Your hardware will become obsolete. Your software will update ad nauseum. Your feeds will continue to scream for your attention by ping, pop-up and click bait headline. But the daily miracle remains unchanged. The account has been credited. The 1,440 minutes are waiting in the vault. The only question that remains is, will you manage the fortune, or will you let it slip through your fingers in a series of unforced errors?

Chapter One Epilogue

Reflect on the daily miracle as a silent, recurring deposit into a 24 hour bank with no rollover and no overdraft protection. If we viewed our time with the same protective scrutiny we apply to our financial assets, we would find it surreal to witness the micro withdrawals we permit every hour to drain our balance. This thought experiment asks you to visualize the 1,440 minutes not as a schedule to be filled, but as a limited fortune of "unmanufactured tissue of the universe" that is either spent on your own life architecture or surrendered to the technical friction of an environment in perpetual flux.

The purpose of auditing this fortune is to recognize the update trap as a form of systemic colonization rather than a series of minor annoyances. Reflect on the time lost relearning an interface or managing the explosion of shadow work produced by high velocity systems. You will identify the technical debt that has been leveraged against your unallocated equity. This is not a task aimed at achieving corporate efficiency. It is an investigation into where your sovereignty has been quietly traded for the illusion of progress through the Jevons Paradox.

Ultimately, this reflection serves as a police report for your

own consciousness. By naming the cognitive residue and the blur problem, you are no longer a passive subject watching your minutes evaporate into a curated stream of digital manipulation. You are choosing to stop waiting for a miracle and instead recognize that you are currently living in one. Are you willing to act as the CEO of your own soul?

Chapter Two

Shadow Work vs. Synthetic Work

Shadow Work, The Harassed Practitioner

THE story of shadow work begins in 1981 with the philosopher Ivan Illich. He coined the term to describe the massive amount of unpaid labor, like cooking, cleaning, and commuting, that was necessary to support the industrial economy but never appeared on a paycheck. To Illich (1981), the shadow was the unpaid darkness that allowed the bright lights of capitalism to shine.

As the digital age arrived, author Craig Lambert (2015) updated this concept. He noticed that technology, which promised to make life easier, actually allowed companies to dump their administrative tasks onto the consumer. Suddenly, we were all travel agents, bank tellers, and grocery baggers, performing self-service tasks that used to be done by paid employees.

However, the most dangerous evolution arrived with a phenomenon we might call the AI paradox. While studies like those by Brynjolfsson (et al. 2023) observe that new tools allow support staff to resolve issues faster, this often does not shorten the work week for the knowledge worker. Instead, it creates an explosion of low-value digital clutter. Shadow work is no longer just unpaid home chores; it is now the paid but hollow labor of the office. It is the exhaustion of the harassed practitioner who spends their day monitoring, editing, and managing a flood of updates, push notifications, optimizations, AI-generated content, and busywork mandates. The cost of producing work fell, so the volume of work expected rose exponentially, trapping the worker in a state of perpetual oversight.

In our context, shadow work is defined as the mandated consumption of cognitive energy on high-volume, low-value tasks, whether unpaid chores or paid administrative busywork loops, that trap the practitioner in a cycle of reactive supervision rather than direct creation.

The Synthetic Work, The Architect of the Soul

Synthetic work comes from a completely different lineage. It is rooted in the esoteric traditions of self-development, most notably the subsidiary exercises of Rudolf Steiner (1904/1994). While shadow work is imposed upon you by the outside world, synthetic work is a choice you make for your own personal enrichment.

Steiner taught that just as a muscle needs exercise to grow strong, the mind needs constructive labor to build mental stature. It is called synthetic because it is an act of synthesis, bringing together scattered parts of your mind to build something stronger.

Synthetic work is the deliberate, internal labor you perform to enrich your own character and *cognitive endurance*. When you practice the Method of Loci, engage in a *shutdown ritual* (both discussed later), or read a complex book to strengthen your reading circuit, you are doing synthetic work. It is the invisible foundation that allows your visible success to stand tall.

While generative AI tools have demonstrated the capacity to increase worker productivity by roughly 14% in specific sectors, the result was not a shorter work week but an explosion of shadow work (Brynjolfsson et al., 2023). This 21st century productivity phenomenon of increased technical efficiency has led to an explosion of additional, often lower value, digital output rather than a reduction in labor. Because the cost of production fell, the volume of output expected by organizations rose dramatically. Practitioners have found their 24 hour living bank accounts drained by the sheer velocity of communication and the new requirements to monitor and edit a flood of busy work. The communication mandates and AI-generated protocols have effectively compressed our human experience into a state of perpetual oversight (reactive as opposed to creative).

To understand the architecture of a modern life, one must distinguish between the two types of labor (Shadow vs Synthetic). Both happen, usually without immediate applause, but they produce opposite results for the human soul. One is a drain that leaves you fatigued; the other is a foundation that makes you strong.

The Work-Type Audit

The transition from just understanding the daily miracle to implementing a functional life architecture is a process. It requires an analysis of the environment in which our 24 hours are spent. In the modern era, the nature of our labor has become the primary determinant of how our time is leaked, taxed, and usurped. To reclaim *intellectual sovereignty*, we must move beyond general time management and perform a rigorous systems analysis of our practitioner/nonpractitioner arrangements. This chapter provides a diagnostic framework for identifying the unique technical friction and cognitive residue generated by the differing on-site, hybrid, and remote work spectrum.

What is intellectual sovereignty in our context? It is the capacity of the mind to function as an independent generator of ideas rather than a reactive processor of external feeds. It is the disciplined refusal to outsource one's curiosity, emotional regulation, or opinion formation to algorithms, group-think, or the feed.

Intellectual sovereignty is the capacity of the individual to insist on self-determination over our cognitive assets. In our digital economy, it is defined as the disciplined maintenance of *cognitive autonomy*. It is the ability to sustain a line of inquiry, regulate emotion, and form judgments. It must be independent of the algorithmic attention capture systems designed to distract human focus and drive the for profit attention economy.

While traditional sovereignty refers to a state's control over territory, intellectual sovereignty refers to the individual's control over their own mental supply chain. It is the refusal to allow external platforms to dictate the salience of information, thereby protecting the mind from the behavioral influences dictated by the attention economy (Wu, 2016).

The Three Pillars of Intellectual Sovereignty

Input Independence: The active curation of information sources to prevent the illusion of independent thought, where a user's worldview is covertly shaped by algorithms rather than conscious choice.

Cognitive Endurance: The ability to resist context switching and sustain deep work or thought for extended periods without interruption. It directly counters the fragmentation of thought inherent in modern interface design (pings and dings) (Newport, 2016).

Emotional Autonomy: The preservation of emotional regulation by disconnecting from the outrage loops of the 24 hour news cycle and social media, often described as a defense against headline stress disorder (Stosny, 2017).

A fundamental error of the 21st century practitioner is the belief that the digital leash is a neutral convenience. In reality, the way we interface with our tools, and the physical locations from which we do so, creates specific vulnerabilities in our 24 hour bank.

On-Site, Threshold and Analog Purge

The onsite practitioner represents the traditional model of labor, yet even this arrangement has been fundamentally altered by the second industrial (technical) revolution. For the individual working in a facility, clinic, or office, the physical boundaries of the building provide a natural structure that the remote practitioner lacks. However, this arrangement is susceptible to a specific form of technical friction, the commute residue.

In 1910, Bennett wrote for the individual taking the suburban train, a journey that provided a clear, physical threshold between the practitioner time and the personal time. Today, that threshold has been colonized. The drive home is no longer a period of reflection; it is often a time when the mind is trapped by cognitive residue. We linger on the mental replay of meetings, the lingering stress of a corrosive interaction, or the recursive thoughts inflicted by a pet project left in a state of incompletion.

The audit for the onsite practitioner begins at the exit. Every minute spent checking one last thing on a smartphone while walking to the parking lot is a micro-withdrawal from the sixteen hours. The primary goal for this work type is to implement an analog purge. This involves a deliberate decision to treat the commute as a practitioner-free firewall. Work related audio or communications are replaced by silence, music, audiobooks, or independent deep thought, effectively closing the practitioner loop before the home threshold is crossed.

The analysis of the onsite practitioner reveals that technical friction is not limited to the digital only sphere. It manifests as a profound weight on those who inhabit physical facilities. The following examples illustrate how technical friction depletes the

24 hour bank of onsite practitioners through interface complexity and administrative colonization.

The Electronic Health Record (EHR) and the Clerical Burden: The modern physician, nurse or allied health field professional serves as the primary case study for technical friction in an onsite environment. Research into the implementation of EHR has demonstrated that these systems, while intended to increase efficiency, have instead turned healthcare providers into data entry clerks. A seminal study conducted by Sinsky et al. (2016) found that for every hour physicians provide direct clinical face time to patients, nearly two additional hours are spent on EHR and desk work. This imbalance leads to a phenomenon often referred to as pajama time; the late-night hours that dedicated practitioners spend at home completing documentation that could not be managed during the shift. This represents a complete failure of life architecture, where the work life threshold is obliterated by administrative debt.

Industrial Noise and Cognitive Performance: The second example of onsite technical friction is found in the industrialized construction sector, where the introduction of intelligent equipment has collided with the physical reality of the job site. The entropy of intelligent systems, when combined with high decibel environments, impose an additional mental workload that prevents workers from efficiently mastering new skills or, more important, maintaining situational awareness. This leads to a state of mental fatigue and cognitive residue where the worker is often unable to turn off the stress of the environment upon leaving the site.

Interface Complexity: The 2017 collision between the USS John S. McCain and a commercial tanker provides us a stark example of how technical friction can lead to catastrophic physical wrecks. As previously mentioned, the National Transportation Safety Board (2019) revealed that the crew was struggling with the highly complex, newly installed touchscreen steering system that replaced traditional and intuitive tactile controls. The interface was so unintuitive that it created significant attention residue. The sailors were so focused on troubleshooting the digital update of their controls that they lost situational awareness of the physical world; the ultimate modern parallel to the 1891 Kipton disaster. Recall, a four-minute watch error lead to a fatal head on collision. Just as that disaster forced the adoption of strict timekeeping standards, the digital errors of the modern age demand new protocols (Boissoneault, 2019). Lessons were not learned. The move to digital mediation was marketed as an upgrade, yet it introduced a friction tax that proved fatal.

The Hybrid Marathon, The Vanishing Wall

The hybrid practitioner occupies the most volatile space on the work type spectrum. Operating in a 3-2 or 4-1 model, alternating between a corporate office and a home environment, this individual is at elevated risk for a deoptimization of life architecture. The hybrid arrangement creates a state of perpetual recalibration, where the mind is forced to adjust to different technical setups, environmental cues, and social expectations multiple times a week.

The primary time leak for the hybrid practitioner is the update trap of the environment itself. Each morning spent reconnecting portable peripherals, troubleshooting new network discrepancies, or relearning an interface that was optimized while out of the office represents technical friction. This work type often suffers from the blur problem, where the 24 hour bank becomes a single, gray marathon of semi work. Without the physical cue of leaving a building, the sixteen hours are easily colonized by practitioner obligations that bleed into, and drain, personal equity.

The audit for the hybrid worker must focus on environmental consistency. The goal is to build digital firewalls that mimic the physical boundaries of an office, ensuring that the work/life blur does not lead to a state of chronic cognitive irritation.

The Remote Employment, The Always-On Era

For the fully remote practitioner, the digital leash is the only remaining structure. This arrangement offers the highest degree of theoretical flexibility, yet it often results in the most severe depletion of the 1,440 minutes. The remote worker is the frontline of the attention economy, operating in a world where the factory whistle has been replaced by the ping of push notifications.

The remote practitioner frequently falls victim to the efficiency paradox and the explosion of shadow work. Because as the cost of communication has fallen, the volume of output expected rises. The remote practitioners spend their day not just doing work but managing oversight of AI content and endless digital communication (pop-ups and pings, feeds and dings).

An audit for the remote worker demands salvaging the concentration core. In the absence of physical walls, the only defense against the algorithm is a disciplined mental muscle. This work type must move from being a user of time to a manager of time, implementing strict technical protocols to ensure that the unmanufactured tissue of their life is not liquidated by the next communication intrusion.

The Unbillable Hour

Whether remote, hybrid, or in the office, many practitioners have been forced to incorporate new layers of digital fatigue. The practice of law is an unshining example of the stolen time required to maintain the digital tools of a trade.

Formatting over Lawyering: In the predigital era, a legal case might involve several boxes of physical paper. Today, a single corporate lawsuit can involve millions of emails, Slack messages, and server logs. To manage this, law firms have adopted complex e-Discovery platforms and AI assisted document review tools. While these tools are designed to increase efficiency, they have introduced a secondary layer of labor; the maintenance of the machine itself (Remus & Levy, 2017).

The Tax, The Unbillable Hour: Research indicates that for every hour an attorney spends on high value cognitive tasks (like legal strategy or trial preparation), they lose significant portions of their day to technical friction. They are troubleshooting software glitches, managing metadata, reformatting documents that broke during a version update, and navigating complex and inefficient court mandated e-filing systems.

The Financial and Human Cost: In the 2023 *Legal Trends Report*, data revealed that while lawyers are working long hours, the average utilization rate remains low at 33%. That means in an eight-hour workday, only about 2.6 hours are billable (Clio, 2023). This represents a direct tax on their expertise. This time is essentially stolen by the machine, as it requires the lawyer to act as an IT technician rather than a legal expert (Susskind, 2019).

The Systems Audit, Quantifying the Friction

To move from diagnosis to resolution, you must now apply these concepts to your own 24 hour bank. This audit is not a test of efficiency but an investigation into the technical debt you are paying every day.

Technical friction and the update trap: Evaluate the time lost to the maintenance of your tools. Estimate the time lost to hardware synchronization, software patches, and general troubleshooting.

Shadow Work and the Oversight Burden: Audit the time spent on lower value digital output. Identify the time spent monitoring or editing automated systems. This is the shadow work produced by the Jevons Paradox. It is effectiveness that has been destroyed by the system (Jevons, 1865).

Cognitive Residue and the Threshold Failure: Measure the psychological weight of your practitioner life during your sixteen hours. Estimate the time lost to the blur problem, where you were semi-working while supposedly at rest.

The Result: Intellectual Debt or Sovereignty?

If your weekly audit reveals a number of anything near 300 minutes lost to these categories, you are carrying a huge time debt. You are effectively paying a technical friction tax of one full work month per year. This is the great social phenomenon that Bennett (1910) noted. It is the person who works only eight hours per day for a livelihood yet finds it difficult to live a full life.

The completion of the work type audit serves as the diagnostic foundation for a much needed restructuring of the 24 hours. Once the technical friction tax has been identified, estimated, and the specific leaks in the daily miracle identified, the practitioner must transition from analysis to construction. This final section of the audit synthesizes the data into a personal life architecture roadmap. It is a structural framework designed to protect unallocated life equity from the systemic pressures of the attention economy (Bennett, 1910; Rosa, 2013).

The primary obstacle for the remote practitioner is the blur problem, where the 24 hour bank becomes an undifferentiated stream of semi-work. Without the physical act of leaving a building, the prefrontal cortex remains in a state of low level arousal, contributing to the Zoom fatigue described by Fauville et al. (2021). This state of perpetual oversight is often punctuated by the third peak phenomenon, where work thoughts spike late at night as the practitioner loses sovereignty over their life. Tasks or thoughts intrude into your personal time by the technical friction created during the standard shift (Microsoft, 2022).

To collapse this peak and protect the sixteen hours, the practitioner must adopt a rigorous shutdown ritual. A successful shutdown ritual is more than a technical logout, it is a psychological deescalation protocol. Cal Newport (2016) suggests that the verbalization of a phrase as simple as "schedule shutdown complete" can act as a definitive cognitive cue to the brain that it is safe to release work related loops. By formalizing this endpoint, the practitioner prevents the attention residue from leaking into dinner, synthetic work or rest (Leroy & Glomb, 2018).

Workplace Telepressure Self-Assessment

This diagnostic tool is designed to measure the tax imposed upon your cognitive endurance. Answer the following questions honestly to determine if you are a *sovereign practitioner* or if you are currently suffering from workplace telepressure.

The Telepressure Audit, Score each statement from 1 (Never) to 5 (Always):

First, I feel an immediate surge of anxiety or a "phantom vibration" when I see a new notification alert.

Second, I check my work email or messaging apps (Slack/Teams) within 15 minutes of waking up.

Third, I need to explain myself or apologize if I don't respond to a nonemergency message within an hour.

Forth, I keep my email inbox or chat window open in a visible tab while trying to perform deep work.

Fifth, I find myself "half-working" (checking messages, etcetera) during dinner, exercise, or time with family.

SCORE:

5–10: **sovereign practitioner**. You have high intellectual sovereignty.

11–18: **Emerging shadow worker**. The update trap is beginning to take hold.

19–25: **High Telepressure**. Your concentration core is under active epistemic capture.

The Cost of the Quick Reply

The primary danger of workplace telepressure is not just the time spent engaged; it is the cognitive switching penalty. Research by Mark et al (2008) indicates that interruptions force a cognitive trade off. While individuals may actually work faster to compensate for lost time, they experience significantly higher levels of stress, frustration, and time pressure.

When a practitioner succumbs to telepressure, responding to five quick messages throughout the morning, they aren't just losing ten minutes of sovereignty; they have lost effective peak cognitive endurance to the shadow work of context switching. This is the mechanism by which time debt is accrued. We borrow focus from our most important work to pay the interest of immediate responsiveness (Barber & Santuzzi, 2015). As we proceed you will learn that nothing is more counter-productive than the always on, always answering mode.

Chapter Three

The Myth of More Time

IF you are reading this book, it is possible you are stressed and tired. It is a specific kind of fatigue that does not come from a lack of effort, but from a surplus of trying. You have likely downloaded the task managers, color-coded your calendar, and adopted the latest methodology for organizing your digital life. You have optimized your workflow, learned keyboard shortcuts to shave seconds off your data entry, and perhaps even listened to podcasts on productivity at one and a half speed. Yet, despite this herculean effort to hack your way to freedom, you feel more behind today than you did five years ago. This is not a personal failure. It is a systemic trap. You have been sold a bill of goods known as the "efficiency myth". It is the belief that if you can just process your tasks faster, automate more, you will eventually clear the decks and find the time to live.

We must begin this chapter with a compassionate truth. The tools are not going to save you. The modern obsession with productivity hacks is not a solution to the problem of time. It is a symptom of our refusal to accept the finiteness of the 24 hours. We have convinced ourselves that with the right app or the right morning routine, we can cheat the physics of the clock. But as we will see, the pursuit of efficiency for its own sake is often the very thing that cannibalizes our self-awareness.

The tragedy of the modern human is best illustrated by the rise and fall of *inbox zero*. The term inbox zero was popularized by productivity expert Merlin Mann in a famous video released in 2007. The philosophy was originally intended to be a system of processing. It was supposed to be a way to clarify what required attention so that one could spend less time in their inbox, and more time doing meaningful work. However, the corporate world stripped the nuance from the philosophy and turned it into a metric. The empty inbox became a badge of corporate virtue, and the full inbox a sign of moral character failure. It was Dilbertism at its best and finest.

Crucially, Mann never intended for the concept to be interpreted as a literal mandate to keep an email folder empty at all times and at all costs. His original philosophy was about processing rather than checking. He envisioned a system to decide the fate of a message (delete, delegate, respond, defer) so that one's mind could leave the inbox and return to higher value work (Google Tech Talks, 2007).

However, the corporate gurus of the time misinterpreted the action (processing) for the goal (zero). You should read *The Peter Principle* at some point. The Peter Principle demonstrates that corporate managers are often promoted until they reach their level of incompetence. Because these leaders lack the skills for their current roles, they compensate by generating excessive administrative friction and shadow work to appear useful. This can create a toxic cycle where incompetent management forces practitioners into low value labor, destroying professional high quality output and organizational efficiency (Peter & Hull, 1969). We have all experienced this to some extent.

The zero became a fetishized metric of control. This corruption of intent led to a scenario where practitioners began prioritizing the administrative task of clearing the queue over the actual work the emails represented. By treating the inbox as a to-do list that must be cleared, the practitioner inadvertently surrenders their intellectual sovereignty to the random priorities of strangers. The empty inbox is not a sign of productivity; it is often a tombstone for the time that could have been spent on deep effective work (Google Tech Talks, 2007; Newport, 2021).

The obsession with inbox zero created a specific psychological phenomenon known as workplace telepressure. Defined by researchers Barber and Santuzzi (2015), telepressure is the unhealthy preoccupation with responding to work related information and communication technology messages quickly. This is not merely a habit; it is a stress response that keeps the body in a state of low level fight or flight.

Barber and Santuzzi's research found that employees who internalized the need for rapid response, a core component of the inbox zero ethos, experienced higher levels of burnout, sleep deficits, and health absenteeism. The pursuit of the empty inbox reinforces the always on expectation, creating a feedback loop where the faster you clear the inbox, the faster others expect you to reply. This destroys the concentration core, as the mind is never allowed to fully disengage from the communication layer to enter a state of deep focus. The zero acts not as a finish line, but as a psychological leash (Barber & Santuzzi, 2015).

Further deconstructing the myth, research indicates that the state of the inbox is less relevant to mental health than the frequency with which we check it. A pivotal study by Kushlev and Dunn (2015) restricted one group of participants to checking email only three times a day (batching), while another group was allowed to check as often as they wished (unlimited).

The results were a direct indictment of the inbox zero compulsion. The group that checked continuously, mimicking the behavior required to maintain an empty inbox, experienced significantly higher daily stress levels. Conversely, the batching group reported lower stress and higher well-being, despite the fact that their inboxes likely contained more unread messages between checks. This empirically proves that the anxiety of the great time delusion comes from the interruption of the workflow, not the volume of the workload (Kushlev & Dunn, 2015). By striving for zero, we invite constant interruption, effectively placing unneeded drains on our unallocated equity in five minute increments. Zero equals stress. Stress drains unallocated equity. As a reminder, our unallocated equity is the approximately eight hours of waking time remaining after practitioner labor and sleep are deducted from the 24 hour bank. Bennett (1910) identified this margin as the "day within a day," representing the only portion of the daily miracle where the individual retains full executive control. In the modern era, this equity is the primary resource targeted for colonization by the attention economy, and by extension the practitioners work efficacy that has been targeted to distract, nag, titillate and profit.

The Empty Vault

Ultimately, inbox zero is a math problem that ignores biology. It assumes that if we clear the inputs, we will find peace. But in an information economy, the inputs are *infinite*. Striving for zero is an unhelpful Sisyphean task, a task that can never be completed, where the boulder rolls back down the hill the moment a new server sync occurs. The empathetic response is to abandon the metric entirely and accept that a full inbox is not a sign of failure, but a sign that you are prioritizing your own output over the world's (and possibly incompetent managers) input.

The result of the zero metric was a collective descent into what author Oliver Burkeman calls the efficiency trap (Burkeman, 2021). By becoming hyper-efficient at processing email, we inadvertently lowered the threshold for receiving it.

Because we responded instantly, our colleagues learned that we were always available, which generated more replies, which required more processing. The tool designed to free us became a feedback loop of unceasing demand. This is a classic manifestation of the Jevons Paradox applied to communication. As we made the transmission of messages easier and faster, we did not gain time. We simply increased the total volume of messages we were required to handle (Jevons, 1865). The practitioner who clears their inbox every hour is not on top of things. They are merely a highly efficient router of information, acting as a switchboard operator for other people's priorities rather than the architect of their own.

This obsession with tools extends beyond communication into the realm of personal knowledge management. In recent years, we have seen an explosion of software promising to build a second brain like Roam Research and Logseq to name a couple. The premise is seductive. If we can just capture every article, every podcast transcript, and every fleeting thought in a searchable database, we will never lose an idea again. We spend hours curating these digital gardens, tagging entries, and linking concepts in a vast web of information. But for many, this practice devolves into what can be called the collector's fallacy.

The collector's fallacy is the false belief that collecting information is the same thing as acquiring knowledge. We feel a dopamine hit when we save an article to a read later app, mistaking the act of archiving for the act of learning. The research on memory and information processing suggests the opposite. A study by Sparrow et al. (2011) on the Google effect (sometimes referred to as digital amnesia) found that when people believe a computer will save information for them, they are less likely to commit that information to their own memory. By outsourcing our intellect to a database, we often clutter our mental environment with administrative tasks; tagging, sorting, filing, while the actual work of deep contemplation remains undone. We build magnificent libraries of unread books and digital files, becoming curators of our own potential rather than creators of actual value. I have been utterly guilty of this. No more.

To understand why these hacks fail, we must look at their historical DNA. Most modern productivity advice is a digital descendant of scientific management, a theory developed by Frederick Winslow Taylor in the late 19th century. Taylor stood on factory floors with a stopwatch, measuring the movements of laborers to eliminate wasted motion and maximize output (Taylor, 1911). This worked brilliantly for moving pig iron or laying bricks where the output was standardized and visible.

The error lies in applying Taylorism to the human mind. We treat our day like a factory floor, breaking our time into fifteen minute blocks and attempting to squeeze maximum efficiency out of every slot. We judge our day by the number of items crossed off a list, regardless of their importance. But the brain is not a machine that operates at a constant RPM; it is a biological organ with rhythms of energy and fatigue. When we use productivity hacks to force the mind into a mechanical pace, we violate its natural laws. We may achieve productivity in the sense of moving data, but we hobble fruitfulness and the ability to generate novel insights and complex work.

As Peter Drucker (1967) famously distinguished, efficiency is doing things right; effectiveness is doing the right things. Our apps make us incredibly efficient at doing things that perhaps should not be done at all.

Take note of the $699 juicer, a case study in efficient futility. To visualize the danger of prioritizing efficiency over effectiveness, we need look no further than the rise and fall of Juicero. They engineered a machine that was a marvel of efficiency. It featured 400 customized proprietary parts, a microprocessor, a scanner to read QR codes, and Wi-Fi connectivity. It applied four tons of force to squeeze proprietary packets of organic produce. By all engineering metrics, the company was doing things right. They had optimized the mechanics of extraction to a level of aerospace precision. However, they failed to ask the question of effectiveness. Was this a problem that needed solving? (Huet & Zaleski, 2017).

The Juicero saga is instructive. The company began as a Silicon Valley fever dream where venture capital giants bet that the humble act of juicing could be turned into a high stakes data subscription. The founder, Doug Evans, painted a picture of a digital garden where a $699, Wi-Fi connected press would serve as the ultimate gatekeeper of health. At the height of the ride, the company attracted over $120 million in venture capital. It was fueled by the marketing promise that this machine offered a superior, high-tech path to wellness that human hands simply could not replicate (Huet & Zaleski, 2017).

The descent was triggered by a single, devastating moment of truth when an investigative report revealed that the expensive "Apple of Juicing" was the definition of irrelevant complexity. A simple video showed that a human could squeeze the juice out of a produce pack with their bare hands just as quickly as the technologically advanced motor, rendering the hundreds of custom engineered parts inside the machine instantly obsolete. As the brand's perceived value evaporated, the company was forced into a desperate price slashing spiral. By September 2017, the roller coaster hit the ground. Juicero liquidated its assets for pennies on the dollar, leaving its investment as a total loss and serving as a permanent warning about the dangers of building a multi-million dollar interface for a problem that was already solved by a human grip (Zaleski et al., 2017).

This was the ultimate exposure of what Lambert (2015) popularized as shadow work. The machine was not adding value; it was adding a friction tax (setup, Wi-Fi troubleshooting, cleaning) to a task, squeezing a bag, that required almost zero effort to begin with.

Juicero serves as the ultimate metaphor for the modern productivity hacker. We are often like that machine. We build elaborate, high-maintenance systems to manage tasks like complex home page dashboards or color coded email filters that could be handled with a simple hand squeeze or, better yet, ignored entirely. We are efficiently processing packets of work that effectively do not need to be processed. We are applying four tons of cognitive force to squeeze a result that yields no nutritional value to our life architecture.

An ingestion pipeline, like Roam Research and Logseq, is a digital system that collects and files information. It can save articles or videos from the internet and organize them for you. While this sounds helpful, it typically leads to the collector's fallacy. This is a mental mistake where you think that just saving a link is the same thing as learning the information.

Because the computer handles the work, you might feel like you are getting smarter. In reality, you are just moving data into an app. You do not actually learn unless you read and think about the material yourself (Forte, 2022). For example, if an artificial intelligence summarizes a marketing protocol for you, you might feel like an expert. But because you never did the hard work of reading it, you still will not understand the topic.

The psychological toll of this constant optimization is the Zeigarnik effect, named after psychologist Bluma Zeigarnik. Her research in the 1920s demonstrated that uncompleted tasks create a state of cognitive tension. They stay active in our memory, nagging at our attention until they are resolved (Zeigarnik, 1927). Modern productivity tools, with their endless lists and reminders, often weaponize this effect. By capturing hundreds of tasks that we cannot possibly complete in a day, we create a permanent state of psychological indebtedness. We look at our to-do list and see not a roadmap, but a catalogue of our failures. This is why you feel exhausted even on days when you accomplish a great deal. The list of what remains undone always outweighs what has been finished.

Reclaiming your life requires a pivot from a philosophy of accumulation to a philosophy of elimination. It requires the empathetic realization that you will never get it all done. The goal of the 24 hour bank is not to fit more into the vault, but to be discerning about what is allowed to enter. We must replace the stopwatch with the compass.

This shift begins with the rejection of time management in favor of energy management. Instead of asking, "How can I fit this meeting into a 15-minute slot?" we must ask, "What will this meeting cost me in cognitive fuel?". Is that meeting even necessary? It means recognizing that a one hour task done in a state of high mental clarity is worth five hours of distracted, “optimized” grinding.

We must also practice the art of strategic incompetence. This is the deliberate choice to be bad at things that do not matter. If responding to email within five minutes is not the primary driver of your life's value, then you must be willing to be the person who takes 24 hours to reply. You must be willing to let the organizational system be imperfect so that the work itself can be excellent.

Ultimately, the myth of more time is a delusion that keeps us running on a treadmill of our own making. Bennett (1910) did not write his treatise so that we could become more efficient gears in the industrial machine. He wrote it so that we could cultivate a wealth of life. The hacks have failed you because they promised to give you control over the uncontrollable. The only true hack is the difficult, empathetic decision to let some fires burn so that you can tend to the one flame that truly matters

.

C-Suite Fable

ONCE upon a time in a Dallas boardroom overlooking the skyline, two executives debated the future of Magic Beans. Miller, a devotee of Excel Dashboards and Pivot Tables, obsessed over the shadow work he deemed essential. He had perfected the bean tracking system, cutting delivery times by half and achieving a frictionless distribution network. "We are doing things right", Miller boasted, pointing to his Inbox Zero Outlook account. To him, the peak of professional life was a perfectly optimized machine, regardless of what the machine actually produced.

Silas, an old hand at consulting who valued intellectual sovereignty, remained silent until the end. "The beans work," he admitted, "but they've grown so fast they've crushed the neighbors' fences and drained the local water supply. You were so efficient at bean counting that you never asked if you should make the beans!"

Miller was the fool of efficiency, polishing a process that was doing things right. Silas was the master of effectiveness. He understood that effectiveness means doing the right things, which includes weighing the human cost of a product. Miller had built a high speed delivery system for disaster. Silas understood that without a concentration core focused on long term impact, speed is just a faster way to cause harm. **The End**.

Fractured Fairy Tales:
THE TALE OF TWO EXECUTIVES
EFFICIENCY
SHADOW WORK
INBOX ZERO
EFFITIENESS
5:00
EFFECTY BEEAMS
Efficiency without Effectiveness
is just Faster Failure!
CH.M

Chapter Four

The Sixteen Hours

WE arrive now at the mathematical heart of our problem. If we strip away the complexities of the work-type audit and the anxieties of the efficiency trap, we are left with a simple, immutable equation. You are allotted 24 hours each day. Physiology demands that you invest approximately eight of these in sleep, a non-negotiable tax paid to the maintenance of your biological machinery. This leaves you with a fortune of sixteen waking hours.

The tragedy of the modern practitioner is not that they lack time. It is that they have fundamentally mismanaged the valuation of these sixteen hours. We have been conditioned to view the eight hours sold to our employers, the workday, as the central event of our existence. The remaining eight hours, the *unallocated equity*, are treated as mere residue, a margin to be spent in recovery, passive consumption, or frantic preparation for the next shift at the factory of life (Bennett, 1910).

This chapter proposes a radical inversion of this hierarchy. We must cease viewing the time outside of work as time off. Instead, we must define it as the *day within a day*; the primary arena where your life is actually lived. Your employer has purchased a subscription to your skills for a limited window. They do not own the copyright to your consciousness. To reclaim your intellectual sovereignty, you must establish a rigorous regulatory framework for this unallocated equity, putting it at the top of the pile, right next to your tax returns.

The Myth of Exhaustion

The primary barrier to reclaiming the sixteen hours is the persistent belief that we are too tired to do anything meaningful after work. We finish our practitioner shift and feel a profound depletion, convincing ourselves that the only remedy is passive rest, streaming content, scrolling social media, or lounging in a state of semi-consciousness.

However, Arnold Bennett (1910) identified this lethargy not as physical fatigue, but as a mental boredom induced by the monotony of the machine. Modern psychology supports this Victorian insight. Research by Sabine Sonnentag (2012) on psychological detachment reveals a counterintuitive truth. Passive recovery (TV, scrolling etc.) is often less effective at reducing stress than active recovery (exercise, social engagement, or learning a new skill).

When we collapse onto the sofa, we are not recharging. We are allowing the cognitive residue of the workday to stagnate in our brains. True restoration requires fresh air for the mind. It requires a shift in focus that displaces the practitioner's worry with a new, engrossing challenge. As we will see, the person who teaches themself French or builds a cabinet in the evening is often fresher the next morning than the person who spent five hours staring at a screen.

Framework, Three Pillars of Protection

To transform the margin into a day within a day, we must build a structure that resists the colonization of the attention economy. We cannot rely on willpower; we need architecture.

The Hard Temporal Firewall

In the corporate world, a meeting room is booked for a specific duration. When the time is up, the meeting ends. Yet, in our private lives, we allow practitioner obligations to bleed into our unallocated equity without a second thought. The first pillar of your new life architecture is the *hard temporal firewall.*

This is a fixed timestamp, let us say 6:00 PM, after which the practitioner identity is formally and totally suppressed. This does not mean you simply stop processing work; it means you enact the shutdown ritual (Newport, 2016). You shut down the software, silence the notifications, and physically leave the workspace. This creates a psychological boundary. It is signaling to your prefrontal cortex that the work tenant has been evicted and the self-tenant has taken possession of the property.

The Sacred Appointment

We rarely miss a doctor's appointment or a client lunch because they are external obligations scheduled on the calendar. Yet we treat our personal aspirations, reading a complex book, writing, or exercising, as optional tasks to be done if there is time. The second pillar is the *sacred appointment.*

You must schedule your unallocated equity with the same rigidity as your practitioner hours. If you intend to read philosophy for forty-five minutes on Tuesday, it must appear on your calendar as a non-negotiable block. If a colleague asks for a late call during that time, the answer is not "I have some free time", but "I have a prior commitment". You need not specify that the commitment is to yourself. By formalizing your leisure, you elevate it from a whim to a mandate (Burkeman, 2021).

Active Detachment Protocols

The final pillar addresses the quality of the time. To prevent the day within a day from becoming a graveyard of good intentions, you must engage in *active detachment.* This connects back to the concept of the concentration core.

Research on recovery experiences indicates that mastery, the experience of learning or improving at something, is a critical component of effective rest (Sonnentag & Fritz, 2007). Your evening activity should offer a moderate level of challenge. This could be the study of technical history, the practice of an instrument, or the repair of an analog device (there are a plethora of options—choose wisely). The difficulty of the task acts as a firewall. It is impossible to ruminate on a passive aggressive email from your boss while you are trying to conjugate a Latin verb or solder a circuit board. The challenge forces the brain to disengage from the work loop, providing true psychological relief.

To illustrate the power of active detachment, we look to individuals who have successfully navigated high-stakes environments by constructing a secondary, intellectually demanding world. These stories demonstrate that the most effective way to recover from the exhaustion of the machine is not to stop using the brain, but to turn its power toward a completely different target.

Marie Curie, The Austerity of Radioactivity

Marie Curie's contributions changed the trajectory of human history, leading to the development of radiotherapy for cancer and the fundamental understanding of the atom. She remains the only person to win two Nobel Prizes in different sciences (Physics and Chemistry). Her labor was the definition of *direct production*, manually processing tons of pitchblende (a radioactive, uranium rich mineral) to isolate a fraction of a gram of radium (E. Curie, 1937).

Curie's sacred appointment was her time at the workbench, which she protected with a ferocity that bordered on the monastic. She famously lived in a state of austerity, ignoring the extreme physical tolls of radiation and poverty to maintain her cognitive autonomy. She and her husband, Pierre, viewed social obligations as a technical friction tax on their research. They refused to even attend the Nobel ceremony initially because it interfered with their laboratory time (Pasachoff, 1996).

Curie's ability to sustain cognitive endurance was legendary. She would work for hours in a drafty shed, her focus so total that she frequently forgot to eat or sleep. This was the sacred appointment as an act of survival. By refusing to patent her discoveries, viewing them as unallocated equity for the benefit of all humanity, she modeled a form of intellectual sovereignty that prioritized the advancement of science over the shadow work of personal wealth management (Quinn, 1995).

Winston Churchill, The Bricklayer and the Artist

Perhaps the most famous example of active detachment is Winston Churchill during the interwar years and the Second World War. When Churchill was not managing the existential crises of the British Empire, he retreated to his home, Chartwell, to engage in two highly specific forms of mastery, oil painting and bricklaying. Churchill famously built a large portion of the kitchen garden walls at Chartwell with his own hands (Churchill, 1948).

Churchill argued that a man can wear out a particular part of his mind by continually using it and tiring it, and that the only remedy was to illuminate another part. For Churchill, painting required a level of visual and spatial concentration that physically crowded out the ruminations of the admiralty. This active detachment provided the fresh air necessary to maintain his concentration core during the darkest hours of the 20th century (Churchill, 1948).

Maya Angelou, The Hotel Room Sanctuary

Maya Angelou's contribution to society extends far beyond her poetry; she was a pivotal voice in the American Civil Rights Movement, serving as the Northern Coordinator for the Southern Christian Leadership Conference at the request of Dr. Martin Luther King Jr. Her work redefined the memoir as a tool for social justice, proving that the personal narrative could reduce systemic prejudice (Gillespie et al., 2008).

To achieve the cognitive endurance required for such emotionally taxing work, Angelou developed a monastic sacred appointment. She famously kept a hotel room, stripped of all personal items, as her primary office. She would arrive at 6:30 AM and stay until 2:00 PM, forbidding the hotel staff from removing any scraps of paper from the floor. This physical isolation was a strategic strike against shadow work. By removing the distractions of home and the telepressure of social expectations, she created a vacuum where only her creative truth could exist (Currey, 2013).

Angelou's appointment was so sacred that she even brought her own dictionary, a Bible, and a deck of cards to tether her mind during moments of fatigue. This ritualistic defense of her attention allowed her to produce work that helped a nation confront its troubled history. Without this rigorous boundary, the deep, rhythmic stature of her prose might have been diluted by the frictionless waste of daily interruptions (Lupton, 1998).

Albert Einstein, The Violinist's Intuition

Albert Einstein is often visualized at a chalkboard. Yet, his most potent tool for active detachment was his violin, which he affectionately named Lina. Einstein did not play to relax in a passive sense. He used music as a technical mechanism to solve stalled intellectual problems. His son, Hans Albert, noted that whenever his father felt he had come to a brick wall in his work, he would take refuge in music, and that would usually resolve all his difficulties (Isaacson, 2007).

This is a classic and profound example of active detachment where the mastery of Mozart or Beethoven acted as a secondary cognitive process. By engaging the auditory and motor cortexes in the complex architecture of a sonata, Einstein allowed his subconscious to process the physics in the background. He not only made the time, but he also defined it. The music was not just an escape from his work; it was a different, nonverbal way of inhabiting his unallocated equity that prevented the blur problem from leading to intellectual stagnation (Isaacson, 2007).

Gertrude Stein, The Architecture of the Modern Mind

Gertrude Stein was the intellectual steward of the lost generation. Her salon at 27 rue de Fleurus in Paris served as the concentration core for the greatest artists of the 20th century, including Picasso and Hemingway. She did not just write; she fundamentally restructured the English language, moving away from 19th century efficiency toward a fragmented, rhythmic style that mirrored the chaos of the modern world (Wagner-Martin, 1995).

Stein's sacred appointment was unique in its brevity and intensity. She believed that it takes a lot of time to be a genius, you have to sit around so much doing nothing (Stein, 1937). Her actual writing often occurred in short, thirty-minute bursts, often while sitting in her Model T Ford while Alice B. Toklas (domestic partner before there were domestic partners) was in a shop. This was not laziness; it was the ultimate expression of effectiveness.

By spending the majority of her day in meditation in motion, walking or observing, she ensured that when she finally reached her sacred appointment, her mental supply chain was fully loaded with insights. She rejected the shadow work of the long workday, proving that thirty minutes of *sovereign focus* is worth more than ten hours of distracted labor. Her contribution was a new way of seeing, forcing the world to look at words as objects rather than just vessels for meaning (Mellow, 1974).

Richard Feynman, The Bongo-Playing Physicist

Nobel laureate Richard Feynman rejected the image of the rigid professor by cultivating a life of intense, varied hobbies, most notably playing the bongo drums and learning to crack safes while working on the Manhattan Project. During his time at Los Alamos, Feynman's obsession with safe cracking was not just a prank; it was a deep, technical study of mechanical systems that provided a sharp contrast to the abstract mathematics of theoretical physics (Feynman & Leighton, 1985).

Feynman's life architecture was built on the principle that the brain remains fresh only if, and if only, it is allowed to play with the unmanufactured tissue of the universe in ways that have no professional utility. His bongo playing and safe cracking were sacred appointments with his own curiosity. This active engagement ensured that even in the highest pressure environment imaginable, the creation of the atomic bomb (with deadlines), his mind remained sovereign and resistant to the update trap of bureaucratic burnout (Feynman & Leighton, 1985).

The CEO of the Soul

Implementing this framework requires a shift in identity. You are no longer a tired employee seeking escape; you are the CEO of a corporation called "my life", which operates primarily between the hours of 6:00 PM and 10:00 PM, and on weekends. The eight hours you sell to your employer are merely a subsidiary operation that funds the parent company.

When you view your sixteen hours through this lens, the time debt begins to clear. You realize that you have not been lacking time; you have been lacking the audacity to claim it. The currency is yours. It is time to spend it on the architecture of a life that belongs to you.

The sacred appointment is more than a calendar invite; it is a defensive fortification for your intellectual sovereignty. Only you can identify the sacred appointment that will bring you value. Once identified from the vast expanse of life's choices, engagement on a consistent basis becomes the grail. It need not be any one thing or a permanent commitment to any one thing. It does need to be a nonnegotiable period of disengagement from the attention economy while offering a challenge to your grey matter.

Your Sacred Appointment

By initialing this sacred appointment Contract, you are acknowledging that your attention is your most valuable capital and that you refuse to let it be drained by the interest of shadow work.

THE SACRED APPOINTMENT CONTRACT

THE PLEDGE

I, as a practitioner of my craft, do hereby declare a state of intellectual sovereignty over my schedule. I recognize that my best work, my direct production, requires a level of cognitive endurance that cannot be maintained while paying the technical friction tax of constant interruptions.

THE APPOINTMENT LOGISTICS

I commit to the following block of uninterrupted labor:
The Hours: ________________ to ________________ (e.g., 8:00 AM – 10:00 AM)
The Frequency: ________________ (e.g., Daily, T/Th, M-F)
The Sanctuary: ________________ (e.g., A specific room, a library, or a disconnected desk)

THE RULES OF ENGAGEMENT

During these hours, I am strictly forbidden from engaging in the following forms of shadow work:
Workplace Telepressure: I will not check email, Slack, Teams, or any real-time messaging platform.
The update trap: I will not perform software updates, system maintenance, or administrative fiddling.
Frictionless waste: I will not scroll feeds, streams or succumb to algorithmic drift.

THE DEFENSE STRATEGY

To protect this concentration core, I have implemented the following checklist barriers:

Hardware: [ ] Phone in Do Not Disturb [ ] Computer disconnected from Wi-Fi [ ] Noise-canceling headphones.

Social: [ ] A Do Not Disturb sign posted [ ] An auto-responder set [ ] Family/colleagues informed.

THE OBLIGATION

I understand that if I break this appointment, I am not just missing a meeting; I am incurring time debt that will be paid for by my future self in the form of stress and diminished stature (Newport, 2016). If the appointment is violated by external forces, I will rebook the time with the same high stakes seriousness I would apply to a tax deadline.

Signed_______________and Dated:________________

The Philosophy of the Appointment

As Peter Drucker (2006) noted, the effective practitioner does not start with their tasks, they start with their time. They do not find time; they take it. By formalizing this commitment, you are a Steward of your own potential. This contract acts as your impetus to stop polishing the shovel and start digging the hole.

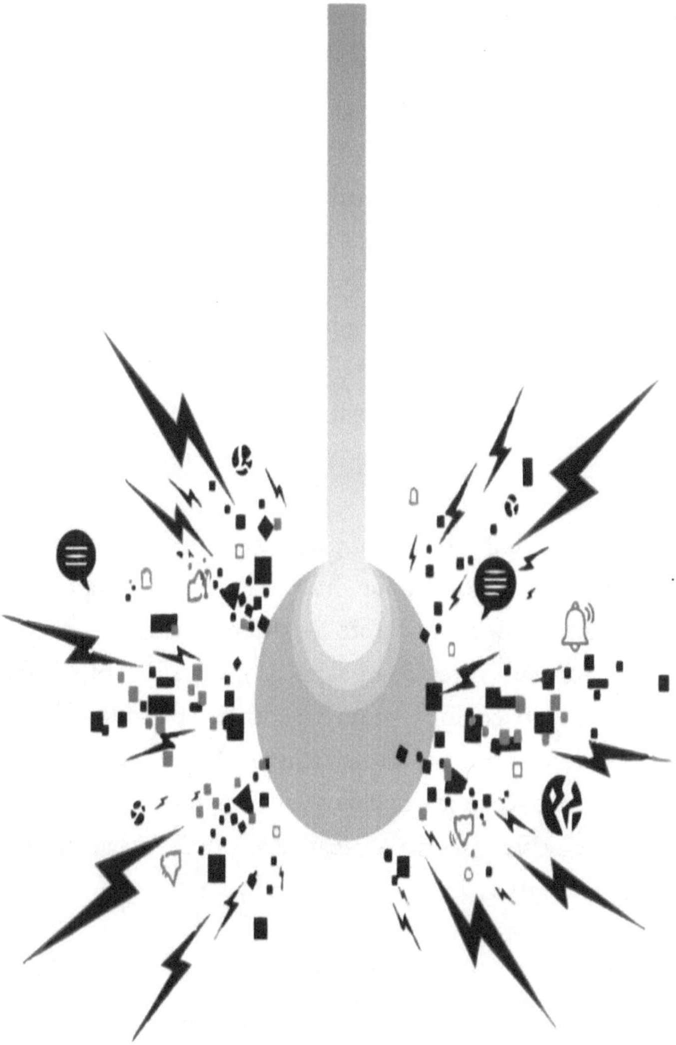

Chapter Five

The Digital Commute

THE construction of a resilient life architecture depends entirely on the integrity of the transitions. In the age of Bennett, the transition was enforced by the geography of the rail line. Today it must be enforced by the geography of the mind. The *digital commute* is not a movement through space. It is a deliberate sequence of technical and cognitive protocols designed to transform the *morning startup* from a period of passive vulnerability into a phase of mental fortification (Bennett, 1910). To reclaim the daily miracle, one must first secure the gates from the barbarians of the morning.

A large serving of caveat prior to proceeding with your personalized blueprint. Your life is not a laboratory with protocols and procedures. Life is often controlled chaos. A single mother with three charges faces a considerably different morning than a bachelor C-suite executive. The following crude guidelines are just that, crude guidelines. Adapt concepts where and as you can. Keep it small to start. It is much better to do a little every day than to do a great deal once a week (Bennett, 1910). Be the CEO.

The Vulnerability of the First Hour

A consistent threat to your intellectual sovereignty occurs in the first hour of the waking day. Most practitioners begin their 1,440 minute daily journey in a state of *digital submission*, reaching for a smartphone or remote before the prefrontal cortex has fully moved beyond the stages of sleep. This act is the ultimate surrender to the attention economy that is, by design, engineered to manipulate you. Fight back. By scrolling news feeds, perusing click bait headlines or digital messages immediately upon waking you are allowing the update trap to set the agenda for your consciousness. You are inviting a crowd of strangers, marketers, and algorithmic ghosts into your life to influence what you should care about before you have even decided who you are for the day.

The brain exists in a uniquely vulnerable state of neuroplasticity during the hypnopompic transition from sleep to wakefulness. While Carr (2010) focuses on the experience dependent plasticity that occurs through repeated digital engagement, recent neurological frameworks suggest that this waking window is defined by a heightened suggestibility as the brain resets its metabolic and synaptic baseline (Tononi & Cirelli, 2014). Engaging willfully with known manipulative digital inputs during this period disrupts the restoration and preservation of the concentration core. By flooding the neural circuitry with algorithmic artifacts before the morning's synaptic pruning is complete, the practitioner triggers a state of cognitive fragmentation that persists throughout the day (Carr, 2010). To resist this reactive posture, one must implement a digital fast until the first *essential engagement* of the day is begun, thereby protecting the mind's sovereign architecture from the update trap.

To qualify as the first essential engagement of the day, a task must be strictly *nonreactive*. It cannot be a response to notifications, headlines, external requests or algorithms that force the brain into a passive state. This engagement must also be an activity that moves the individual towards long term goals such as exercise, journaling, meditation or lunch prep to name a few. The key is a period of analog input/output for the first waking moments of your day. Finally, this engagement must be predigital, occurring during the specific neurological window when the brain is still completing its natural synaptic pruning from the night before. That requires the individual to complete the task entirely before their first digital log-in to protect their mind's sovereign architecture from the update trap. A five minute period of meditative day planning is better than a digital screen plastered with headlines, click-bait and marketing pitches.

The Architecture of the Morning Startup

The startup of the practitioner's machine of obligations must be a secondary event to the startup of the human soul. Modern life architecture suggests a *threshold protocol* that mimics the physical commute of the 19th century, a period devoid of the attention economy, allows the mind to expand into personal priorities before it is compressed into a practitioner's role (Bennett, 1910).

This protocol consists of three essential phases:

The Analog Buffer (The first twenty minutes): This is the period of essential digital silence. The smartphone is limited to the function of an alarm clock. The goal here is *cognitive priming*, engaging the mind with an analog, high-value source. Whether it is reading, silent reflection or just mentally blueprinting the days goals, this phase ensures that the first occupant of your mind is of your own choosing. The point is to select an analog task. A nonfiction book on the nightstand, analog chore or a meditative routine is an excellent go-to.

The Strategic Oversight (bridge from analog to digital): Before engaging with the incoming queue, you must review the outgoing mission. This is where you consult your life architecture roadmap. You are acting as the CEO of your soul, verifying the nonnegotiable blocks of your sixteen hours before the practitioner subscription begins. This could be mulling the days to-do list while eating a bagel. It could be reminding themselves how they disengage from their practitioner role eight hours from now as you brush your teeth. It can be as simple as planning your *controlled launch*. The first two protocols may often blend into one. Be the CEO. Try to place a force field of analog protection between you and the attention economy for those first few moments.

The Controlled Launch (Engage): Only after the first two phases are complete do you cross the digital threshold. This is a deliberate log-on rather than a passive drift-in. You open tools with a specific intent, utilizing a ready-to-resume plan possibly created the previous evening or during the analog buffer to avoid the switch-cost tax of relearning where you left off (Leroy & Glomb, 2018). Specific intent is the key here. Choose, preferably the night before, a specific task to complete prior to subjecting

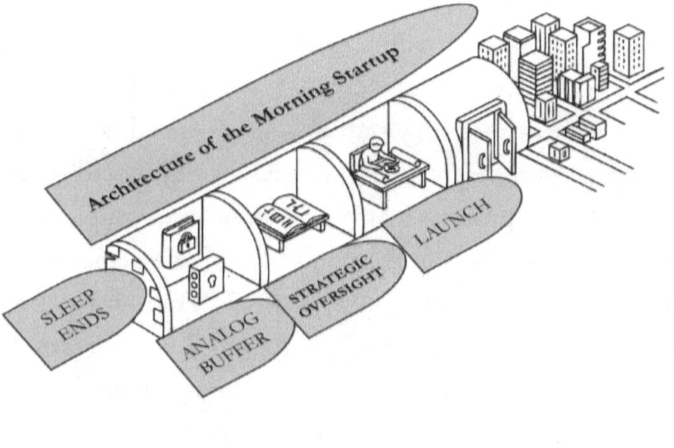

yourself to being manipulated by algorithms, pings, feeds or external demands on your time.

To ensure that the practitioner identity is formally suppressed and the personal soul is liberated, we can look to practitioners in high consequence fields who have mastered the art of the threshold protocol. These individuals recognize that without a deliberate ceremony of transition, the work tenant never truly vacates the mind, leading to a state of permanent cognitive depletion.

Toni Morrison, Capturing the Pre-Secular Light

Toni Morrison's contributions to the global literary world are monumental. She was the first African American woman to win the Nobel Prize in Literature (1993) and a recipient of the Pulitzer Prize for *Beloved* (1988) (Gillespie, 2008). Her work fundamentally reshaped American literature by centering the black experience with a prose style that combined epic scale with intimate, psychological depth. However, for much of her career, she was also a single mother and a senior editor at Random House, roles saturated with the shadow work of household management and the administrative friction of the publishing industry (Gillespie, 2008).

Morrison's morning startup was designed to reclaim her intellectual sovereignty before these roles could claim her attention. Her process was built around the pre-secular space; the time before the sun rises and the world's demands become secular or ordinary. She would rise at 5:00 AM, make coffee, and sit as the light slowly entered her room. This ritual was not just about time; it was about transition. She noted that observing the dawn allowed her to access a specific, non-secular space where she was prepared to inhabit her characters' lives (Schappell & Brodsky Lacour, 1993).

By writing on yellow legal pads starting around 5:00 AM, Morrison ensured that her direct production was completed while her cognitive endurance was at its peak. She avoided the update trap of the modern era by refusing to engage with any external information until her morning's work was done. This architectural choice allowed her to produce eleven novels while maintaining a demanding professional career, proving that a rigorously defended startup is an excellent way to sustain stature in a distracted world (Currey, 2013; Schappell & Brodsky Lacour, 1993).

Benjamin Franklin, The Scheme of Moral Perfection

Benjamin Franklin remains one of history's most productive polymaths (broad knowledge). His accomplishments range from inventing the lightning rod and bifocals to serving as a primary Founding Father and a world renowned diplomat. His ability to move between the physical labor of a print shop and the high stakes direct production of scientific and political theory was made possible by a literal architecture of the day, which he detailed in his autobiography as his "scheme" (Franklin, 1791/1996).

Franklin's startup began at 5:00 AM and was centered on an audit of his unallocated equity. His process was a three-hour block dedicated to rising, washing, and addressing "powerful goodness"! This was followed by a non-negotiable directive; "prosecute the present study" (continue a project). By placing his most difficult intellectual labor in the 5:00 AM to 8:00 AM window, he ensured that his intellectual sovereignty was established before his business as a printer could impose its technical friction tax.

This startup allowed him to ask the foundational question of his day; what good shall I do this day? This was not a vague hope but a strategic goal for effectiveness, doing the right things rather than just doing things efficiently. Franklin's morning architecture allowed him to cultivate the cognitive endurance necessary to master multiple languages, conduct complex electrical experiments, and help draft foundational documents like the U.S. Constitution. All this while managing a profitable business (Franklin, 1791/1996; Isaacson, 2003).

Ernest Hemingway, The First Light Protocol

Ernest Hemingway's impact on 20th century literature was so profound that he was awarded the Nobel Prize in 1954, largely for his mastery of the art of narrative (Mellow, 1992). He famously pioneered the Iceberg Theory, a prose style that relied on extreme cognitive autonomy to strip away unnecessary waste, leaving only the essential truth of the story. Achieving this level of clarity required a morning startup that was as disciplined as a military operation (Mellow, 1992).

Hemingway's process involved rising at first light, usually around 6:00 AM, and working until noon. He famously wrote in a standing position, which acted as a physical hard temporal firewall. When his legs grew tired, he knew his cognitive endurance was nearing its limit for the day. His architecture was built on the principle of stopping while still productive; that is, stopping when he still knew what would happen next. This ensured that his sacred appointment for the following morning would have an immediate on ramp, preventing the update trap of staring at a blank page (Plimpton, 1958).

By completing his direct production before lunch, Hemingway protected his work from the shadow work of the literary life, the letters, the interviews, and the social posturing of the afternoon. This protocol allowed him to produce nearly 1,000 words a day of high-quality, publishable prose. His morning architecture was not a choice; it was a survival strategy for his talent, ensuring that the world only got what was left over after his craft had taken its share (Mellow, 1992; Plimpton, 1958).

The Surgical Ready-to-Resume Ritual

High stature surgical teams should utilize a highly structured handoff and shutdown ritual that serves as a vital mental threshold. This protocol acts as a clean transfer of information, ensuring that the patient's critical data, such as hemodynamic stability and intraoperative complications, is not lost to the shadow work of administrative drift. By formalizing this process, the team protects the concentration core of the recovery staff, preventing the cognitive fragmentation that occurs when a high stakes task is left in a state of unresolved ambiguity (Haynes et al., 2009; Starmer et al., 2014). Rituals prevent errors.

In real-world application, however, the implementation of these rituals is frequently stymied by technical friction and time-pressure constraints (Peter Principle?). While organizations like the World Health Organization mandate these shutdowns, global compliance is uneven, with structured handoffs occurring in only 30% to 60% of clinical transitions (Starmer et al., 2014; WHO, 2009). Often, the trap of a busy operating schedule forces teams into a reactive posture, where they bypass the formal audit in favor of a hurried, informal exchange. This failure of direct communication creates a potential of error that correlates directly with increased perioperative mortality. It proves that when the ritual is sacrificed for speed, the human tax is paid in patient safety (Haynes et al., 2009). Execution has been inconsistent but where applied, it pays dividends in enhanced patient outcomes.

This protocol mimics the ready-to-resume plan identified in organizational science (brief, intentional plan you make before switching away from a task). By explicitly documenting the status of a patient and handing over cognitive ownership to the on-call team, the surgeon closes the open loops in their prefrontal cortex. This ritual prevents having the Zeigarnik Effect, where uncompleted tasks are remembered more readily than completed ones, from following them home. This ensures that the daily miracle of their sixteen hours is not spent in a state of ruminative anxiety (Leroy & Glomb, 2018; Zeigarnik, 1927).

The Analog Bridge, Naval Officers, the Tactical Purge

Naval officers aboard modern destroyers, such as those navigating the complexities of the touchscreen driven USS John S. McCain, face the extreme risks of cognitive overload. The National Transportation Safety Board (2019) found that complex digital interfaces can contribute to a loss of situational awareness and a degradation of decision making capabilities.

To counter this update trap, a tactical purge becomes a necessary survival strategy. This requires a transition from the digital mediation of the screen to an analog verification of the physical horizon. By physically stepping out onto the bridge wing to verify the ship's position, the officers create a sensory threshold. This act grounds the brain in the physical world, stripping away the technical friction of the touchscreen and allowing for a clean mental transition. It is a deliberate reclamation of the concentration core through analog grounding (National Transportation Safety Board, 2019).

The Digital Fast of the Creative Practitioner

In the realm of high velocity creative work, author and computer scientist Cal Newport (2016) documents the protocol used by theoretical physicists and writers to transition into deep focus. One specific example is the monastic threshold ritual. An individual maintains a physical workspace that contains no digital inputs. No Wi-Fi, no smartphone, and no computer with internet access.

This protocol acts as a *digital commute*. By physically moving from a connected office to a disconnected library or study, the practitioner triggers a neurological shift. The act of passing through the doorway into a digitally silent room serves as a hard firewall. This threshold ensures that the manipulative attention economy cannot follow the worker into the anticipated deep study. It is a physical manifestation of the sacred appointment, where the environment enforces the integrity of the sixteen hours (Newport, 2016).

The Physical Commute as a Mental Laboratory

For those who traverse physical space to reach a work environment the journey is often viewed as wasted time to be endured. This is a miscalculation each practitioner should rethink. The commute is the sacred appointment of the transition. Bennett (1910) marveled at the practitioner who traveled forty minutes each way and yet had no time for study. He argued that the train was a private laboratory for the mind.

In the modern context, we should resist the audio colonization of the commute. The digital manipulation by algorithms. The temptation to listen to news, professional podcasts, or productivity hacks is often just another form of the efficiency trap. It keeps the mind in a state of semi-work. It hinders true psychological detachment. To fortify the mind, the commute should be used for *deep integration* that results from a neural rewiring process.

If you drive, this may mean periods of silence to process complex ideas or directed self-talk to resolve a technical problem. If you take public transit, it means the rejection of the smartphone in favor of a physical book or a notebook. The goal is to arrive at the workplace not as a tired traveler, but as a fortified intellect who has already accomplished the most important work of the day; the management of oneself.

Reclaiming the Transition

The digital commute is the firewall that protects the daily miracle from the erosion of the always on culture. By treating the startup and the transition as technical ceremonies, you ensure that the practitioner cog does not overwrite the human governor.

If you fall short of managing these thresholds, you will find yourself paying an ongoing technical friction tax. You will arrive at your sixteen hours fatigued not by your work, but by the friction of failing to leave it behind. The morning is the vault. The commute can be the combination to the padlock. In the next chapter, we will detail the specific mind control exercises required to strengthen the concentration core and ensure that once these thresholds are established, they remain impenetrable to the manipulations of the algorithm.

Thought Experiment, The Iceberg Protocol

Goal: Transform a 500-word shadow work narrative into a 50-word direct production masterpiece using the Theory of Omission.

Hemingway's Theory of Omission, or the Iceberg Theory, is a strategy for cutting out the technical friction of boring, unnecessary details to reveal the concentration core of a story (Hemingway, 1932/1999). It works by hiding seven-eighths of the actual information underwater, leaving only the most important details visible on the surface. Less is more in this context. This style forces you to do the direct production of thinking for yourself instead of just falling into an update trap where the author explains every single little thing (Mellow, 1992).

To use this life architecture, a writer must have the cognitive autonomy to delete anything that feels like fluff. By keeping the writing simple and sharp, you create a clean transfer of feeling that hits the reader much harder than a long, wordy essay. Hemingway proved that a story actually gains more impact when you have the courage to leave out the parts you know are there but don't need to say (Hemingway, 1932/1999).

Step 1 The Bloated Draft: (The shadow work version) Read the following text. Notice how the excessive detail creates technical friction, making it harder to feel the actual emotion of the scene.

He sat there in the chair, a wooden chair that felt hard against his back, and he looked out the window at the rain that was falling down in heavy, grey sheets against the glass pane that was slightly cracked in the corner. He felt a deep sense of sadness, a sort of heavy melancholy that seemed to weigh down his shoulders like a leaden coat. He remembered the way she used to smile, that specific curve of her lips that always made him feel like everything was going to be alright, but now she was gone, and the house felt empty and hollow, like a drum that no one was beating anymore. He picked up his coffee cup, which was cold now, the steam having long ago vanished into the stagnant air of the kitchen, and he took a sip that tasted bitter and metallic, much like the way his life felt at this very moment in time. He wondered if he should call her, or if he should just sit there and let the shadows grow longer across the linoleum floor, which was stained with a spot of spilled wine from the night before when they had argued about something so trivial he couldn't even recall the details now. The silence was loud, a ringing in his ears that wouldn't stop, and he felt the urgent, desperate need to do something, anything, to break the stillness, yet his limbs felt like they were made of stone, unmoving and heavy, pinned to the chair by the sheer gravity of his own regret. He thought about the letter in the drawer, the one he hadn't opened yet, with the postmark from a city he had never visited, and he felt a shiver of fear, or maybe it was anticipation, crawling up his spine like a thousand tiny insects with cold, prickly feet...

Step 2 The Direct Production: (The 50-Word Solution) Now, apply the Iceberg Theory (Plimpton, 1958). What is the 7/8ths that we can omit? We don't need to describe the sadness; we need to show the state of the man. Trust the reader to feel the underwater mass of the story.

The coffee was cold. Rain streaked the cracked pane, grey and rhythmic. He looked at the unopened letter from a city he did not know. In the kitchen, the silence was heavy enough to hold him in his chair. He did not call. He just watched the shadows stretch

Chapter Six

The Concentration Core

WE have now established the architecture of the day. We have built the firewalls. We defined the thresholds. We audited the friction. Yet a perfect system is less than optimized if the operator lacks the strength to run it. You may carve out the sixteen hours, but if your mind remains a puppy in the park, bouncing about with every push notification, impulse, and shiny digital object, you will find yourself floundering in the middle of your unallocated equity. You can be paralyzed by the inability to choose wisely.

Arnold Bennett (1910) identified this point of weakness over a century ago. He observed that while people could perform complex practitioner tasks under the duress of management, they lacked the mind control to direct their own intellect during their leisure time. Today, this atrophy is not really a lack of discipline; it is a neurological injury. We have spent decades outsourcing our memory to Google and our sense of direction to GPS. The concentration core has atrophied from disuse imposed upon us by profit motivated manipulation. We know sugary snacks and tobacco have negative health benefits. Less obvious has been the algorithmic misuse of humanity, until now.

To reclaim the daily miracle, we must engage in a regimen of mental kinetics. This concept, developed by designer Michael R. Gill, operates like a focus fortress for creators (Gill, n.d.). An instance of this application was Gill's role as the lead designer of the *Army History Journal*. His design for the publication featured

Douglas E. Nash Sr.'s article, *Kesternich: The Battle That Saved the Bulge* (Nash, 2018). The brutal fighting for the German village of Kesternich in December 1944 played a decisive, often overlooked role in shaping the outcome of the Battle of the Bulge. Gill elevated the historical narrative through a visual architecture that matched the rigor of the scholarship. This achievement underscores the difference between efficiency (making a magazine look professional) and effectiveness (creating a vessel that honors the weight of the content).

We must treat the brain not as a hard drive to be filled, but as a muscle to be expanded. We must pump mental iron. This chapter details the specific exercises required to rebuild your cognitive attention span, moving you from a state of algorithmic reaction to a state of intellectual sovereignty.

The Physiology of Attention

Before we begin the exercises, we must understand the mechanism of failure. Modern neuroscience distinguishes between bottom-up attention (the reflexive focus triggered by a loud noise to a subtle ping) and top-down attention (voluntary focus directed by the prefrontal cortex) (Carr, 2010).

The attention economy is engineered to hijack your bottom-up system, typically with a profit motivation at its core. Every breaking news, every vibrating alert, and every infinite scroll is designed to bypass your executive function. When you feel tired after a day of scrolling, it is because your bottom-up system has been firing rapidly, depleting your neurochemical reservoir. Building the concentration core is the process of strengthening the top down control circuit, the brain's command center, primarily located in the prefrontal cortex. It is the ability to say, I will look at this, and only this. It is the ability to hold that gaze despite the screaming protests of the dopamine receptors (Newport, 2016).

Exercise I: The Bennett Protocol

The first task is deceptively simple and notoriously difficult. It is derived directly from Bennett's original prescription for mental discipline.

The Directive: Select a mundane, physical object. A tchotchke, a watch, or a chair. For a period of ten minutes, you must focus your concentration upon this object. You may study or research its manufacture, its history, its materials, or its design, but you must not allow your mind to wander to your email, your dinner plans, or your anxieties.

When your mind inevitably drifts (and it will, likely within thirty seconds), you must gently but firmly drag it back to the object. This act of dragging back is the mental equivalent of a bicep curl. It is the friction that builds the muscle. Neurological data confirms that *as little as eight weeks of consistent practice* can physically reorganize the brain's architecture, creating a stable, nonreactive foundation for all subsequent sovereign thought (Hölzel et al., 2011). Bottom line? This can actually raise your IQ (Mrazek et al., 2013).

To operationalize the Bennett protocol, we need not look at theoretical models, but at documented historical cases where the deliberate concentration on a mundane object was used to engineer a superior mindset. The following two examples illustrate that the concentration core is not a gift, but a forged instrument.

The Steiner Pencil Ritual

In the early 20th century, philosopher Rudolf Steiner developed a series of subsidiary exercises designed to strengthen the cognitive faculties of his students. The first and most critical of these was the Control of Thought exercise (Steiner, 1904/1947).

Steiner argued that the modern mind was like a leaf in the wind, (puppy in the park) blown about by associative thoughts and sensory distractions. To cure this, he prescribed a rigorous daily protocol. The student must select a simple, uninteresting man-made object, specifically suggesting a pencil, a paper clip, or a safety pin. For five minutes daily, the student was required to focus their entire mental attention exclusively on this object.

The rule was strict. The mind could not wander to personal memories associated with the object (e.g., my grandmother had a pencil like this). It had to remain fixed on the object's physical reality. Fixation was on its manufacture, its materials, and its form. If the mind drifted, it had to be *forcibly* brought back. This exercise was not geared toward learning about pencils. It was about the force used to hold the image. By treating a boring object as the center of the universe for 300 seconds, the student built the mental muscle required for higher-order thinking. "To be free is to be capable of thinking one's own thoughts — not the thoughts merely of the body, or of society, but thoughts generated by one's deepest, most original, most essential and spiritual self, one's individuality." (Steiner, 1904/1947)

The Flaubert-Maupassant Tree Observation

The second example comes from the literary world. It demonstrates how this protocol creates cognitive persistence. In the 1870s, the young writer Guy de Maupassant sought the mentorship of the master realist Gustave Flaubert. Flaubert did not teach him plot or grammar. He taught him what would become the Bennett protocol of vision.

Flaubert would walk Maupassant to a tree and issued a command. Describe this tree to me so that it cannot be confused with any other tree. "The left limb rises dutifully, branching in the expected, symmetrical way. The right limb juts out horizontally for nearly a yard before curling upward, as if it had been trying to point at something specific". He insisted that Maupassant stare at the object, a cab horse, a man smoking a pipe, or a blazing fire, until he discovered a detail that no other person had ever seen.

"There is a part of everything which is unexplored... To describe a blazing fire or a tree in a plain, we must remain before that fire or that tree until they no longer resemble for us any other tree or any other fire." (Maupassant, 1888)

This was not a creative writing exercise. It was a torture test of attention. Maupassant was forced to push past the simple scanning by the brain which just says, that is a tree, and engage in deep, sustained focus. The focus was pursued until the reality of the object revealed itself. This training is credited with developing Maupassant's legendary observational powers. Proof that the concentration core produces not just discipline, but sometimes genius (Maupassant, 1888).

To paraphrase Bennet; the mind is a servant. It can be trained to concentrate on the most humdrum subject. When it wanders, it can be brought back. To quote Bennett, "Bring it (your mind) back by the scruff of the neck… (In ten minutes)you will have brought it back about forty times". (Bennett, 1910) Discipline transforms drift.

Practice this daily, possibly during your analog buffer phase (Chapter Five). This is not about the kettle, the tree, the cab, or the cat. It is about the assertion of your inherent authority. You are proving to your own consciousness that you are the captain of the ship.

Exercise II: The Deep Reading Sprint

Cognitive resistance can strike back at the algorithmic death star. The internet has trained us to read in an F-pattern, scanning headlines and bullet points while ignoring the body text (Nielsen, 2006). This scanning behavior makes it physically uncomfortable to engage with dense, linear arguments. To reverse this, we must practice deep reading sprints.

The Directive: Select a text that is slightly above your current comfort level, a philosophical essay, a technical manual, or complex fiction. Joyce's *Ulysses* or Faulkner's *The Sound of Fury* come to mind. Set a timer for twenty minutes. During this window, you must read linearly. No skipping! No skimming! No checking definitions on a secondary device. Struggle with the context.

If you encounter a difficult passage, do not page away. You re-read it until it yields its meaning. This exercise trains cognitive persistence. Learn, train and grow the ability to stay with a problem when the initial dopamine hit fades. Maryanne Wolf (2018) argues that this deep reading process is essential for critical thinking and nurturing empathy. By refusing to skim, you are rewiring the neural pathways that allow for sustained thought. Again, you are raising your IQ.

Bill Gates, The Think Week Protocol

The most prominent modern example of institutionalized deep reading is Bill Gates's think week. Twice a year, arguably the busiest man in the technology sector would retreat to a secluded cottage in the Pacific Northwest for a week. He disconnected completely from his daily operational duties to engage in a marathon of deep reading.

Gates did not use this time to skim emails or read business summaries. He consumed dozens of complex technical papers, PhD dissertations, and lengthy industry proposals (Guth, 2005). No passive consumption. It was active interrogation of the text. Gates would write extensive margin notes and formulate entirely new strategic directions for Microsoft based on this linear and uninterrupted processing of dense information.

This practice illustrates the core tenet of the deep reading sprint—cognitive persistence. By removing the digital inputs that fragment attention, Gates forced his brain to engage with level 3 complexity (*synthesizing* new ideas) rather than staying trapped in level 1 processing (*reacting* to data). The *Internet Tidal Wave Memo*, which steered Microsoft toward the web in 1995, was a direct result of one of these deep reading sprints (Guth, 2005).

The *Internet Tidal Wave Memo*, authored by Bill Gates and distributed on May 26, 1995, is one of the most significant strategic pivots in the history of technology industries. It serves as a definitive example of an organization recognizing a fundamental shift in effectiveness. Microsoft then performed a massive, top-down course correction to avoid being swept away by a change in the global landscape (Gates, 1995).

The Context, A Focalizing Shift

In early 1995, Microsoft was a powerhouse with the impending release of Windows 95. However, the company was primarily focused on MSN (The Microsoft Network) as a proprietary, closed system. It was a form of interface sovereignty that ignored the open, chaotic growth of the interwebs.

Gates's memo was a radical act of cognitive autonomy. He realized that the internet was not just a feature but a fundamental shift in the technical system of human communication. He famously declared that the internet was the most important single development to come along since the IBM PC was introduced in 1981 (Gates, 1995).

The Core Directives

The memo was structured to force a state of direct production across every department of the company. Gates identified several critical success factors that redefined Microsoft's life architecture roadmap.

The Browser as the Shell: He recognized that the web browser would become the primary concentration core for users, potentially making the operating system (Windows) irrelevant if Microsoft did not control the interface.

Averting technical friction: Gates admitted that competitors like Netscape were moving faster and with more stature. He ordered that every product group, from Office to Windows, make the Internet their number one priority (Cusumano & Yoffie, 1998; Gates, 1995).

The Emperor's New Peel Moment: He stripped away the illusion that Microsoft's current success guaranteed future survival. He used the memo to dismantle the shadow work of internal bureaucracy and refocus the entire workforce on the coming digital tsunami.

The Result: The browser war. This memo triggered a massive reallocation of unallocated equity. Microsoft pivoted almost overnight, bundling Internet Explorer with Windows 95 and effectively crushing Netscape's market share in what became known as the Browser Wars.

While this move eventually led to the landmark U.S. v. Microsoft antitrust case, it successfully integrated the internet into the core of the personal computer experience. For the sovereign practitioner, this memo serves as a masterclass in how to identify a tidal wave of change and how to use a hard temporal firewall (a nonnegotiable, time-based boundary) to stop old habits and start a new era of production (Cusumano & Yoffie, 1998; Guth, 2005).

Kodak, The Barbarians at the Gate

Imagine if Eastman Kodak had such a leader in the 1990s. George Eastman was that kind of leader or there never would have been an Eastman Kodak Company that sprung roots in the 1880s.

George Eastman bypassed his lack of schooling by standing in Chicago bookstores and reading entire encyclopedias until he mastered the concentration core (American Business History, 2018; Profectus Magazine, 2023). He devoured technical photography journals to gain intellectual sovereignty over the topic (American Heritage, 1972; Bennett, 1910). This deep reading sprint was essential. It prevented any dilution of purpose in his future chemical experiments (PBS, 2026). Discipline created industry!

The 1990s seemingly gave us the opposite. There was a digital visionary within the corridors of Kodak. The architect of the first digital camera was Steven Sasson. He was an engineer who presented his prototype to Kodak leadership in 1975. Sasson's device was a toaster sized assembly that utilized a Charge-Coupled Device (CCD) to capture black-and-white images with a resolution of 0.01 megapixels. This innovation represented a clean transfer of photography into the digital realm, yet executive leadership viewed it as a technical friction tax on their chemical film empire. Innovation was stifled.

Sasson predicted that digital innovation would achieve market leadership over film within twenty years. That timeline proved startlingly accurate. However Kodak's leadership chose to remain within an update trap of legacy success. Vision was silenced. This failure to engage in direct production around the digital prototype allowed competitors like Sony and Nikon to seize the market.

Kodak's purported institutional paralysis serves as the ultimate demonstration of how an obsession with a legacy product, in this case analog film, created a lethal business death spiral. It blocked a transfer to emerging digital architectures (Lucas & Goh, 2009). Allegedly, the executive board perceived the first digital prototype as an attack on existing margins rather than the future of the industry (Gavetti et al., 2005). While holding the intellectual sovereignty of the original digital patents, they prioritized chemical processing over digital reality. Perceived safety became suicide. Tens of thousands of Rochester, NY citizens paid the price in lost prosperity. The 2012 bankruptcy instantly dissolved $1.2 billion in retiree health and survivor benefits for 56,000 citizens. George Eastman most likely would not have allowed a cadre of executives to facilitate such destruction. If Kodak Execs had learned from the example their deep reading founder (and Bill Gates) had set, then Kodak might still be a household name.

Several Kodak executives reportedly danced away with multiple millions of dollars in salaries and/or consulting contracts while the Greater Rochester Area citizenry suffered financial destruction.

Maryanne Wolf, The Rehabilitation of the Reading Brain

A more poignant example comes from cognitive neuroscientist Maryanne Wolf. Despite being an expert on the reading brain, Wolf discovered that her own concentration core had atrophied due to years of digital scanning and screen-based distractions.

When she attempted to read Hermann Hesse's *The Glass Bead Game*, a complex, philosophical novel she had loved in her youth, she found it physically impossible. Her eyes skimmed over the page, her patience snapped. She could not force her brain to slow down enough to comprehend the dense prose (Wolf, 2018).

Wolf realized she had fallen victim to the F-pattern scanning habit. To cure this, she did not simply try harder. She implemented a deep reading sprint protocol. She forced herself to read the text for fixed periods (twenty minutes), resisting the urge to skim, until her brain's deep reading circuitry re-engaged. It took two weeks of this painful, deliberate friction before she regained the cognitive patience required to enjoy the book. Two weeks is a very small price to pay.

This example serves as a clinical validation of the exercise. It proves that the inability to focus is not a permanent character flaw, but a reversible neurological condition that has been perpetrated by algorithmic profit mongers (perhaps often unknowingly). Through the resistance training of the deep reading sprint, the brain can be retaught to inhabit the sixteen hours with depth (Wolf, 2018).

Think of your life as a castle. The hard temporal firewall is the heavy gate you pull shut every night to keep the noise of the world out. No emails, no digital chores, no homework after a certain hour. But sometimes, a storm hits that is big enough so that it forces you to open that gate. When this happens, you do not just let the storm flood the whole castle; you handle it like a pro.

The first thing you have to do is stop and look closely at the problem. Is this a real emergency that only you can fix, or is it just the fake urgency of a 21st century communication making you feel like you should be busy? If it is just a distraction wearing a mask, keep the gate shut. But if the situation really needs you, you must prepare to step into the breach with a plan (Perlow, 2012).

Once you decide to help, set a containment window. This is like lighting a candle that only burns for thirty minutes (you can still get a 30 minute hourglass; old school personified). You tell yourself that you will only work until that candle goes out. This keeps the emergency from stealing your whole night and turning your time into shadow work, that draining empty work that doesn't actually get you anywhere (Lambert, 2015). You go in, do exactly what needs to be done, and *ignore* everything else. It is a surgical strike. Fix the problem and get out before the digital junk pulls you back in.

Ending the work is just as important as starting it. When your timer goes off, you have to sanitize your mind. Log exactly where you stopped so your brain does not keep worrying about it while you're trying to sleep (shutdown protocol). Use active detachment. It is the mental lock that lets you walk away and actually feel like you're done for the day (Sonnentag & Fritz, 2015). By declaring the breach closed, you give yourself permission to rest.

Next morning, during your morning startup, you look back at what happened. Do not be mean to yourself about it. Look at the facts. Why did the storm get through the gate? Was it because of a technical glitch, or because you did not set a clear enough boundary with a friend or a boss? By figuring out what went wrong, you can make your castle even stronger for the future (Drucker, 2006). Turn a stressful moment into a lesson in intellectual sovereignty, proving that you are the boss of your own time, the CEO.

Exercise III: Slay the Google Effect

The *google effect* (digital amnesia) suggests that we no longer remember information because we know we can retrieve it instantly (Sparrow et al., 2011). This leaves the mind empty of the raw materials required for creativity. You cannot connect dots that you do not hold in your head.

The Directive: Commit a specific piece of trivial information to memory every week (not hard to find on this planet). This could be a poem, a sequence of historical dates, or the taxonomy of a local plant species.

This is not about becoming a trivia master. It is about resisting the dependency on the external brain. When you memorize a stanza of poetry, you are building a mental fortress. You are declaring that your mind is a place where things can dwell, not just a corridor through which data passes. This internal archive becomes the furniture of your sixteen hours, making your own mind a more interesting place to inhabit (Bennett, 1910).

To prove that the human mind is not a fixed hard drive but a dynamic biological vessel capable of expansion, we look to two definitive case studies. These examples illustrate that when we resist the google effect and commit complex systems to the *internal archive*, our brain physically changes to accommodate the new equity.

Joshua Foer, The Journalist Who Forgot to Be Forgetful

In 2005, science journalist Joshua Foer attended the U.S. Memory Championship to observe, believing that the competitors were savants with freakish biological gifts. He soon discovered that they were ordinary individuals who had simply trained their internal archive using the Method of Loci.

The Method of Loci is an ancient craft that transforms your mind into a vast, familiar estate. Knowledge is no longer a fleeting ghost but a permanent resident. To build this memory palace, you mentally traverse a place you know by heart, a childhood home or the corridors and rooms of a previous school. Anchor complex ideas to specific landmarks along the way. By turning abstract data into vivid, strange images and placing them on your mental shelves, you move from the shadow work of mindless repetition to a state of intellectual sovereignty (Yates, 1966). When you need to recall a fact, simply walk through your mental rooms, seeing each truth exactly where you left it. It will prove that a well ordered mind is the ultimate fortress against the frictionless waste of digital life (Foer, 2011).

Foer, who described his own memory as average, spent one year (just one year) training his brain to resist the externalization of data. He did not use an app. He used the ancient technique of spatial visualization to encode decks of cards and random numbers into the architecture of his mind. In 2006, he returned to the competition not as a reporter, but as a contestant, and won the U.S. Memory Championship. He set a new record for speed cards (memorize the order of a shuffled 52-card deck) (Foer, 2011).

His journey demonstrates that the google effect is a choice, not a destiny. By building an internal library of images, Foer proved that the average professional could expand their working memory capacity. We can transform the mind from a sieve into a fortress (Foer, 2011).

The London Taxi Drivers & The Hippocampus

Perhaps the most potent biological argument for the internal archive is the study of London taxi drivers. To earn a license, a driver had to pass The Knowledge. It is a grueling test requiring the memorization of 25,000 streets and thousands of places of interest within a six-mile radius of Charing Cross. The process takes an average of three to four years of deep, daily study. In a landmark longitudinal (years long) study, researchers followed a group of trainees before and after they attempted the exam. The study found that those who successfully qualified and thus built a massive internal archive of London streets exhibited a significant increase in the volume of gray matter in their posterior hippocampus (the region associated with spatial

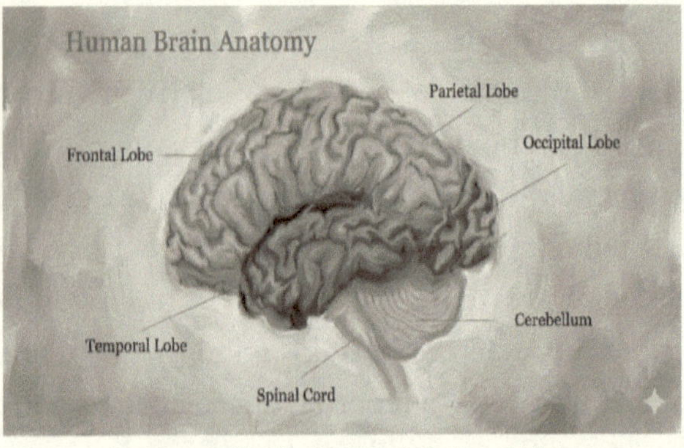

Image created by Gemini from author's multiple prompts.

memory). Those who failed or dropped out exhibited no such growth (Woollett & Maguire, 2011).

This confirms that the adult brain remains plastic. When we refuse to count on GPS (the ultimate google effect tool) and instead force the brain to remember the route, the biology of the brain literally expands to meet the demand. We do not just learn the information; we become the vessel for it and the keeper of it.

The Result, A Mind That Obeys

The goal of the concentration core is not to turn you into a robot. It is to give you the power of choice. When the core is weak, you do not choose your leisure. The willfully manipulative algorithm chooses it for you. When the core is strong you have options. You can stand in the middle of a chaotic digital landscape and choose to look at a flower, read a book, or simply think a thought through to its conclusion. You have now built the time. You have fortified the transition. You have trained the muscle. Next chapter? We will discuss the ultimate purpose of this architecture. Not to produce more work, but to cultivate the high stakes leisure that makes your 24 hours' worth living.

Chapter Seven

The Reflective Mood

WE have built the walls. We have trained the muscle. We have audited the friction. Now, we must inhabit the house.

The concentration core is the engine that drives your attention. The *reflective mood* is the atmosphere in which that engine operates. It is the emotional and intellectual climate of a life reclaimed from the manipulative algorithm. Arnold Bennett (1910) argued that the ultimate goal of managing one's time was not efficiency, but a state of being he called the reflective mood. It is a consistent, low-humming background of philosophical inquiry that colors the entire day. Its importance cannot be overstated.

In our modern 21st century environment, the reflective mood is the antithesis of the *reactive mood*. The reactive mood is the default state of the digital citizen; a perpetual, high-frequency scanning for the next input, the next crisis, the next dopamine hit. It is a life lived in the accusative case, always being done to, rarely doing. It is a subtle but omnipresent overlord.

The reflective mood, by contrast, is a life lived in the nominative. It is the quiet confidence of a mind that knows it is the author of its own experience.

The Philosophy of the Gap

The foundation of the reflective mood is the cultivation of the *gap*. Austrian neurologist and holocaust survivor Viktor Frankl famously attributed his survival, his sanity, to the realization of what he termed the last of the human freedoms. While popular culture often summarizes this as a space between stimulus and response, Frankl's actual words cut deeper. "Everything can be taken from a man but one thing; the last of the human freedoms — to choose one's attitude in any given set of circumstances, to choose one's own way" (Frankl, 1946/2006)

Imagine being in a place so cold and terrifying that you feel like you have lost everything. Gone, seemingly forever, is your home, your family, and even your own name, replaced by a number. This was the world of Viktor Frankl. He was a doctor held as a prisoner in concentration camps during World War II. In a place where people tried to treat him like he did not matter, Frankl discovered a powerful secret that helped him stay alive. He realized that even when he was treated with utter cruelty, no one could reach inside his mind to change how he felt. He observed that while we cannot control the scenario (the stimulus), we retain absolute sovereignty over our reaction. In that internal fortress, we are the boss of ourselves, and we have the power to choose who we want to be (Frankl, 1946/2006).

While many around him succumbed to despair, Frankl survived by focusing on his why. His reason for living. He spent his time thinking about his wife and the important work he wanted to finish once the war was over. He would even close his eyes and imagine himself standing in a warm, bright classroom, teaching students about the lessons he was learning in the camp (Pattakos & Dundon, 2017). By keeping his eyes on a goal for the future, he found strength to keep walking through the darkest days. He proved that even when we cannot control what happens to us, we always get to decide how to respond.

In the digital age, technology is intentionally designed to collapse this gap. The notification ping (stimulus) demands an immediate swipe (response). The email arrives(you've got mail!), and we reply. The headline flashes. We feel outrage. The reflective mood is the deliberate, intentional widening of this gap. It is the discipline of pausing for five seconds before opening the app. It is the refusal to have an opinion on a news story until 24 hours have passed. It is the decision to let the phone ring.

Widen the gap and we move from being a switchboard operator to being an architect. We create a buffer zone, a safe zone, where wisdom can take root.

The Practice of Daily Review

To maintain this stoic mood, we must structure and perform a closing ceremony for the sixteen hours on a consistent basis. Just as we have a morning startup to fortify the mind, we need an *evening review* to settle it. This is not a to-do list for tomorrow; it is an audit of the soul for today.

The Stoics, particularly Seneca, practiced a nightly examination of conscience in his treatise *On Anger*. Scholars paraphrase the original Latin as; "When the light has been removed and my wife has fallen silent... I examine my entire day and go back over what I have done and said, hiding nothing from myself, passing nothing by" (Seneca, 2010, 3.36.3–4).

For the modern practitioner, this practice is the antidote to the Zeigarnik effect (remembering unfinished tasks better than finished tasks)(Zeigarnik, 1927). Review the day, acknowledge the victories, accept the failures. Note the moments where the concentration core held firm and we close the loops. We tell the brain, the day is done. The data is filed. You may rest. Without this review, we carry the friction of Tuesday into the dawn of Wednesday, adding to our cranial fatigue.

The Return to Cause and Effect

The ultimate symptom of the reactive life is the feeling that things just happen to us. The weekend disappeared. The evening got away from me. The language of a victim is not inspirational or uplifting. The reflective mood enhances the understanding of cause and effect.

When you live in a reflective mood, you understand that a lost evening is not a mystery; it is the result of a failed threshold protocol. You understand that your exhaustion is not a curse. It is the logical physiological cost of a triple peak day (shadow work bursts morning, afternoon and evening). This is not about guilt. It is about data. When you understand the mechanics of your own life, you stop hoping for more time and start engineering it.

You become a student of your own existence, observing your energy levels like a biologist observing a specimen. You treat your failures not as moral indictments, but as system errors to be patched in the next update.

To demonstrate that the reflective mood is not merely a philosophical ideal but an actual tool for high performance, we examine two modern figures who have utilized rigorous daily audits to structure their lives. One uses data to track cognitive output, while the other uses ancient philosophy to track moral character.

Jim Collins, The Quality Spreadsheet

Jim Collins, the celebrated author of *Good to Great*, engineers his own life with the same empirical analysis he applies to Fortune 500 companies (Collins, 2001). Collins has maintained a sophisticated daily tracking system for decades. He employs a spreadsheet where he logs his creative hours (time spent on deep thought and writing) and records a daily quality score.

At the end of every day, Collins briefly reviews his days performance and assigns a grade; +2 (a great day), to -2 (a bad day). He then calculates a running average to ensure his life is trending in the right direction. This is not only record keeping; it is a daily review that creates a feedback loop. By reviewing the data, he can identify exactly which activities (or people) cause a minus 2 day and systematically eliminate them. He doubles down on the factors that produce a plus 2 day (Ferriss, 2019).

This practice perfectly illustrates the CEO of the soul concept. Collins does not just hope for a good life; he engineers it by auditing the data of his own existence daily.

Ryan Holiday, The Stoic Evening Audit

Ryan Holiday, a modern proponent of Stoic philosophy and author of *Stillness is the Key*, has engaged in the nightly routine of Seneca for the 21st century. Holiday performs a specific journaling protocol each evening to close the gap between his actions and his values.

Instead of a freeform diary, Holiday's review focuses on specific questions designed to interrogate his personal actions, logging wins and shortcomings. Questions such as; "Who do I spend time with? What is the most important thing? Is this who I want to be?" This is an active detachment ritual that forces him to confront his own behavior while the memory is still fresh.

Holiday argues that his nightly audit is the only way to prevent the drift of modern life, where one's character is slowly eroded by the friction of the attention economy (Holiday, 2019).

By externalizing the day onto paper, Holiday clears the cognitive residue before sleep, ensuring that the successes and failures of Tuesday do not haunt the morning of Wednesday.

The Final Sovereign

The goal of this book is absolutely *not* to make you a more productive worker. The world has enough of those! The goal *is* to make you a more fully realized human being. The reflective mood is the mindset in which you inhabit your unallocated equity.

It is the moment you stand on the train platform, and, instead of checking your email, you check the sky and follow the path of a bird. It is the time you spent struggling with a difficult text, not because it will make you money, but because it will make you sharper. It is the realization that the daily miracle is not a resource to be toiled and mined, but a gift to be unwrapped.

The 24 hours are yours. The vault is unlocked. The friction has received lubricant.

You are the only one who can decide how to spend the currency of your life.

Suggestions for Ending Your Day

Look back through the hours to find exactly where your time was stolen by the pull of a screen or the mindless drift of an algorithm. **Catch the thief.** Admit exactly where you lost your way or chose an easier path. **Name the mistake.** Separate the noise of the day from the one moment where your effort was truly commensurate with your highest goals. **Keep the gold.** Ask if your mind was a servant today or a master. **Rule your thoughts.** Forge a single, sharp directive for tomorrow's concentration core. **Light the path.**

Chapter Eight

Deep Reading in a Shallow Age

IF the concentration core is the turbocharged engine of your new life and living architecture, then the book or preferably the tome, is the fuel. If you have gotten this far without skimming you are well on your way. To quote a Snoopy thought bubble you, "deserve a pat on the head by a human being". (strangerkar, n.d.).

We must make a critical distinction before proceeding. When we speak of reading in this chapter, we are not referring to the consumption of information. You read thousands of words a day. You skim emails, Slack messages, captions, and tweets. This is decoding. It is a rapid-fire, superficial processing of symbols to extract data. It is efficient, necessary, and physiologically shallow.

Deep reading is an entirely different can of worms. It is the sustained, immersive engagement with a complex and linear narrative or argument. It is consciously surrendering your internal monologue to another voice, of another human, for a set period of time. It is the textual equivalent of active listening. In the age of the manipulative algorithm, this act has become counter cultural. It is a torches and ropes rebellion against a world that demands we skip to the end.

The Death of the Linear Mind

Since the nineties, we have been participating in a worldwide, uncontrolled experiment in neuroplasticity (Carr 2010). In his seminal work *The Shallows*, Carr identified that the internet is not just a tool, it is a medium that demands a specific type of cognition. The digital medium prioritizes speed, multitasking, and the F-pattern scan. The F-pattern, where the eye darts across the top of a page and down the left side, is hunting for keywords (Nielsen, 2006).

Carr's (2010) analysis argues effectively that the internet is a technology of neuroplasticity, actively rewiring the neural pathways of its users. Far from being a neutral delivery system, the medium itself encourages a style of thinking that is rapid, distracted, and fragmented. Little wonder that Al Gore declined taking credit for the internet. The brain, adapting to the environment of the web, has become highly proficient at skimming surface level data. However, it progressively loses the cognitive endurance required for deep, linear reading (Carr, 2010).

This cognitive shift is physically observable in user behavior. Nielsen's (2006) eye-tracking studies revealed that web users rarely read content word for word. The employed F-Pattern scan is as thus:

Top Horizontal: First, the eye sweeps across the upper part of the content area (the headline or first paragraph).

Middle Horizontal: Then, the eye moves down and sweeps across a second, shorter horizontal line.

Vertical Scan: Last, the eye scans down the left-hand side of the page in a vertical movement, hunting for key words or bullet points.

This F-shaped consumption pattern changes the user from a contemplative reader into a hunter gatherer reader. While efficient for finding specific data points, this mode of input effectively sacrifices nuance and retention, reinforcing the brain's preference for the reactive mood over the reflective mood (Carr, 2010; Nielsen, 2006).

The human cost of this adaptation is the deoptimization of the *linear mind.* The linear mind is patient. It can hold a premise in Chapter One, suspend judgment through Chapter Three, and formulate the conclusion in Chapter Ten. It is capable of nuance, ambiguity, and deferred gratification.

When we stop reading deeply, the ability to finish a difficult and dense book is hampered. We lose the ability to think in long chains of logic. We become susceptible to the soundbite, the gaslight, and the hot take. Many of our brains have lost the stamina required to unpack complex truths.

The Biliterate Brain

Cognitive neuroscientist Maryanne Wolf (2018) warns that we are in danger of damaging the reading circuit. That is a specific network of neurons that connects vision, language, and cognition. Different from speech, which is genetic, reading is an invention. The brain must be rewired and trained to do it. If we stop exercising this circuit, it willingly submits to entropy and atrophies.

We are seeing the ubiquitous emergence of the skim reader, brains that processes everything, even Shakespeare or a legal contract, as if it were a Reddit thread. The skim reader misses beauty, misses subtext, and most critically, misses empathy. Wolf (2018) argues that deep reading is our primary simulator for empathy; it forces us to inhabit the consciousness of another person more fully. Without it, we retreat into the silos of the self. Fewer examples of an empathy desert are more on point than our emotional keelhauling by Mr. 45 and Mr. 47. Take note of the evolution from bone spurs (Eder & Philipps, 2018) to "losers and suckers" (Goldberg, 2020). When did empathy become a mortal sin in a "Christian" nation?

Wolf (2018) explains that there is no single reading center in the mind. Instead, the brain has to forge a brand new connection between the parts that see shapes, the parts that understand language, and the parts that think deep thoughts. This reading circuit is a plastic, changeable network. If you exercise it with complex books it becomes a passing lane on a superhighway for critical thinking.

However, this circuit is use it or lose it. Wolf (2018) warns that as we spend more time skimming social media feeds and scanning headlines, we are retraining this circuit to be shorter, shallower and faster. That bypasses the deep, slow parts of the brain responsible for logic and understanding other people's feelings (empathy). The war on woke is nothing more than a war on empathy. We are trading our ability to understand complex arguments for the ability to process simple data quickly. We are embracing a future where we can read words but can no longer grasp their deeper meaning. We are metaphorically and psychologically burning books.

Susceptibility to Demagoguery

If you are not concerned it matters little where you are on the intellectual spectrum. Wolf (2018) warns that the skimming brain processes information too rapidly to perform the slow, essential and demanding work of critical analysis. When a population loses the patience to follow complex arguments or verify sources, they become biologically and emotionally vulnerable to demagoguery.

The result? Take careful note. A citizenry that cannot distinguish between a reasoned argument and an emotionally charged slogan is a willing slave, embracing and encouraging the language of the despot wannabe. The deep reading circuit is the same circuit used for critical thinking. Its atrophy leaves the public unable to detect false equivalencies or manipulative rhetoric. This is a target rich environment for authoritarian leaders who traffic in gaslighting and simple, violent lies.

The Collapse of Cognitive Empathy

Deep reading, specifically of dense fiction is one of the few activities that forces the brain to simulate the consciousness of another person. Wolf (2018) argues that without this gymnasium of empathy, we retreat into tribalism.

The result is a society where individuals are physically incapable of perspective-taking. Instead of doing the difficult work of imagining the other, the skimming brain retreats to familiar silos of information that confirm its existing biases. We become willing subjects of group think and echo chambers. We are all guilty of this on some point on the spectrum. This leads to a dangerous dehumanization of anyone outside one's immediate tribe, as the neural capacity to feel the reality of a stranger's life has withered away into hate, or worse.

To reclaim the daily miracle, we must cultivate a biliterate brain. A brain that can skim the screen when necessary but can shift gears into the deep, slow first gear of the page when it matters (Wolf, 2018).

The Tome as an Anchor

In a life of fluid notifications and tribal buffeting, the physical book serves as a status anchor. It is finite. It does not change. It does not vibrate. It does not watch you. It does not judge you.

When you sit down with a physical book during your unallocated equity, you are engaging in a tactile and physical protocol of disconnection. The medium itself forces the concentration core to engage. You cannot tab away from a printed page. Read it or close it. That binary choice destroys the illusion of multitasking (and multitasking is an illusion). When an individual endeavors to multitask, performing two tasks that require conscious direct production, the brain does not operate in parallel. It executes a rapid, sequential alternating thought process known as task-switching (Levitin, 2014). Simultaneity is an illusion.

Furthermore, a book represents the ultimate Jevons Paradox reversal. It is inefficient. It takes a long time. It *cannot* be optimized. That is the point of the exercise. By choosing to do something slow in a fast world, you reassert your sovereignty over time (Jevons, 1865). You declare that your attention is too valuable to be chopped into six second segments by the manipulative algorithms, protocols and feeds.

The Protocol of Active Engagement

To truly read in a shallow age, we must move beyond passive consumption. We must treat reading as a contact sport. This involves the resurrection of marginalia (margin notes).

Samuel Taylor Coleridge

Coleridge was famous for his marginalia; writing notes, arguments, and insults in the margins of the books he read. This is the highest form of respect one can pay to an author. It signals that you are not just receiving the text but wrestling with it.

Coleridge was perhaps the world's worst houseguest, but the world's greatest reader. He had a notorious habit of borrowing books from friends, most often the essayist Charles Lamb, and keeping them for months, or even years. However, his friends rarely complained. Why? Because Coleridge returned the books enriched with usury. According to Lamb, the intellectual interest added by Coleridge's pen made the volumes more valuable than they were in their original state. (Lamb, 1823).

Coleridge did not just read a book; he moved into it. He treated the white margins of the page not as empty space, but as a silent room where he could debate with the author. He was so prolific in this practice that he is credited with popularizing the very term marginalia to describe these notes.

Here are two authentic instances of his marginalia that capture the spectrum of his mind:

The Humorous: When confronted with words from an early poem he coauthored alongside Robert Southey, Coleridge wrote in the margin, "Hang me, if I know or ever did know the meaning of them" (Lezard, 2003). Audit your own ghosts.

The Poignant: Coleridge's most cited and deeply moving marginalia note occurs in his 1823 audit of Leighton's *Commentary on the First Epistle of Peter*, where he writes:

"I have no hope. I am a poor, helpless, crest-fallen, and almost heartbroken wretch. I have no stay within myself, no support out of myself. I am a reed shaken with the wind" (Jackson, 1980, p. 601).

John Adams

This habit of talking back to the page was the core of his daily structure (Ellis, 1993). He treated every volume as a chance for a direct trade of ideas, testing the author's high-flown theories against his own gritty experience with human nature.

John Adams viewed Jean-Jacques Rousseau's high-flown theories about natural freedom as a complete fallacy that ignored how humans actually start their lives (Haraszti, 1952). While Rousseau was busy writing that everyone is born free, Adams was in the margins of his book, essentially rolling his eyes and providing a reality check (Adams, n.d.). He pointed out the obvious defect between theory and fact, scribbling that a human is actually born as a helpless infant who depends entirely on a nurse or parents just to survive (Haraszti, 1952, p. 81). Dependency is reality on day one. By calling out this fallacy, Adams used his innate insight to show that freedom is not some magical gift we are born with, but something much more complicated that we have to work for (Ellis, 1993). Reality trumps romanticism.

Externalize Your Thoughts Onto the Page

One should rarely ever read deep text without a pencil in hand. Underline the arguments. Argue with the premises. Build your own index at the back of the book (or separate page with referenced page numbers for the book conservatist).

When you externalize your thoughts onto the page, you transform the book from a monologue into a dialogue. You stop being a consumer. You become a collaborator. This physical interaction with a book anchors the information in your internal archive, impeding the google effect from washing it away with the next ping (Sparrow et al., 2011).

Exercise Your Brain

To test if you are truly reading or just scanning, grab a non-fiction book or a complex novel. Open to a random page or start a new chapter. The rules are simple. You are not allowed to turn two pages until you have logged exactly three specific things in the margin.

First: Find and underline (or transcribe) one sentence that reminds you of something that actually happened in your own life. In the margin, write a brief note like This is similar to when I... or my version of this is... By connecting the author's words to your own memory, you anchor that fact in your mind so you are less likely to forget it.

Second: Locate one sentence that you disagree with, or one that seems confusing. In the margin, write a challenge. Is this true? But what about... or I don't buy this. If you are not arguing with the author, you are merely accepting data. Test the ideas to see if they are strong enough to stand on their own.

Third: Circle the most complicated paragraph on the page. In the margin, consolidate that entire paragraph into a five word summary of your own. If you cannot explain it simply, you do not understand it yet.

If your page looks messy when you are done, you win. A clean page is the sign of a disengaged mind.

The Sanctuary of the Page

End this chapter with a challenge. The world is algorithmically screaming for your attention. It creates panic to sell you ads. It creates outrage to keep you scrolling. A book is the exit door.

When you open a book, you step out of the frenetic now and into the eternal. You can commune and debate with minds that have been dust for millennia. You access the accumulated wisdom of the species, unfiltered, untainted by an algorithm that prioritizes engagement over truth. Resist. Be the CEO of your own brain.

The Protocol of Active Engagement, Modern Case Studies

To illustrate that active engagement (specifically the use of marginalia and nonlinear processing) is not an antiquated habit but a vital tool for modern intellectual production, we look to two contemporary information architects who have built careers on their ability to wrestle with texts.

Ryan Holiday, The Churn Protocol

Ryan Holiday, author of *The Obstacle Is the Way*, utilizes a rigorous, multi-stage protocol for active engagement that he learned from his mentor, Robert Greene. Holiday does not treat reading as a passive consumption event, but as a mining operation.

His protocol involves distinct steps of engagement. He reads the physical book with a pen, aggressively marking passages and folding the corners of pages that contain critical insights. He employs a digestive gap; he places the finished book on a shelf for a few days or weeks to allow the information to settle. Finally, he returns to the book to process the marginalia and dog eared pages. If the marked passage still resonates, he manually transcribes it onto a 4x6 index card, which is then filed into a physical commonplace book organized by theme (e.g., Strategy, Death, Ego) (Holiday, 2017).

This physical friction, writing, waiting, transcribing, and filing, ensures that the information is not just seen by his eyes but is encoded into memory. Holiday credits this analog active engagement as the engine for his prolific output. He stated that by the time he sits down to write a book, the reading has already done the work for him.

Maria Popova, The Alternative Indexing Method

Maria Popova, the creator of *The Marginalian* (formerly *Brain Pickings*), reads hundreds of books a year, yet she rejects the speed reading philosophy in favor of deep, structural engagement. Her protocol is creating a shadow index within the book itself.

Popova does not merely highlight; she builds her own taxonomy at the back of the book. She lists page numbers corresponding to specific acronyms she has invented, such as BL for beautiful language or G for grief. This effectively rewrites the book's content to match her own intellectual landscape. She is not relying on the author's structure. She is actively imposing her own architecture onto the text (Ferriss, 2014).

This method transforms the book from a static object into a dynamic database. When she later condenses these ideas for her publication, she is not searching for an unreliable memory; she is retrieving a specific data point from a map she drew herself. This is the ultimate expression of the biliterate brain, using analog tools to organize vast amounts of information with a depth that algorithms cannot replicate.

Deep reading is not a hobby. It is an essential protocol of hygiene for the soul. It is the only way to ensure that in a world of shallow waters, and competing distractions, your mind retains its depth.

Chapter Nine

The Amateur's Renaissance

Danger, Danger, Will Robinson

WE have arrived at a dangerous intersection in modern life. We have successfully carved out the sixteen hours and built the concentration core. The temptation, fueled by a culture obsessed with optimization, is to ask, "How can I monetize this?"

If you spend your evenings crafting from your inspirations, you are urged to open an Etsy shop. If you create prose, you are told to build a Substack following. We have been conditioned from adolescence to believe that an activity is only valuable if it generates capital or attracts attention. This chapter argues for the *amateur's renaissance.* To reclaim the word amateur in its truest sense, it is the person who engages in a craft for the love of the process and the result.

The Monetization Fallacy

The side hustle is the enemy of the day within a day. When you monetize your leisure, you introduce the logic of the marketplace into the monastery of your soul. By turning your unallocated equity into a part-time job, you destroy the psychological detachment you studied and worked so hard to build. You are no longer detaching from the machine; you have simply built a smaller machine in your home office, shop or kitchen. However, accidents do happen. Unintended consequences; cave actionum improvisos eventus tuarum for the Latin fans.

John Plant, Primitive Technology

In 2015, John Plant began walking into the bush of Far North Queensland with zero tools. In complete silence, he filmed himself engaging in synthetic work; building tiled roof huts and firing clay kilns using only sticks, stones, and mud. Plant did not do this out of necessity; he did it for the pure, inefficient mastery of the process.

The authenticity of his labor struck a nerve in a digital world. As of February 2026, his YouTube channel, *Primitive Technology*, maintains over 11 million subscribers (Plant, n.d.). The project, intended only as a personal study of ancient mechanics, eventually resulted in a publishing deal (Plant, 2019). By rejecting the efficiency of the modern world, Plant created something the modern world valued, the sight of a human being doing real and ancient work.

The Rise of Synthetic Work

As we move deeper into the age of automation, we face a paradox. Machines are increasingly capable of performing the necessary work of survival. The Jevons Paradox suggests that as efficiency increases, we often consume more rather than work less (Jevons, 1865).

But there is a higher path called synthetic work. As established in previous chapters, synthetic work is effort that is objectively unnecessary but subjectively vital. It is the deliberate choice to do things the hard way to maintain human engagement with biological and physical reality.

The Protocol of the Grand Hobby

To enter the amateur's renaissance, you must select a *grand hobby* that is difficult, productive, and fulfilling. Winston Churchill (1948) sought refuge from the pressures of leadership by painting and laying bricks at his estate. While his invitation to the bricklayers union was later contested and became a point of political friction, his commitment to the manual craft remained essential to his mental stability (Best, 2001).

Similarly, Emily Dickinson (1955) produced nearly 1,800 poems in private manuscripts, refusing to seek a public audience. Richard Feynman and Ralph Leighton (1985) recount, as previously discussed, how Feynman played the bongos, seeking a tactile rhythm far removed from the abstractions of theoretical physics.

Nick Offerman, The Woodshop Antidote

Actor Nick Offerman describes manual labor as a highly constructive medicinal necessity. He argues that the profession of acting is inherently abstract, and woodworking produced the corrective friction. Nick Offerman has often spoken about the importance of hands-on work, woodworking, and craftsmanship in his life, especially in connection with his role as Ron Swanson in the sitcom *Parks and Recreation.* (Offerman, 2023) This is a primary example of synthetic work. The value of the table is not the furniture itself, but the psychological grounding the act of making it provides to the maker.

The Final Freedom

When you engage in synthetic work, you step out of the hierarchy of competence. You are free to be bad at it. This freedom is the ultimate luxury of the sixteen hours. By becoming a serious amateur, you remind yourself that you are not just a unit of economic production. You are a creator.

Chapter Ten

Curiosity as a Competitive Edge

WE have spent the previous chapters mining for and reclaiming time, building focus, and establishing the philosophical architecture of the day. We now must turn our attention to the specific utility of the mind you have liberated. In the industrial era, the very valuable worker was one who could answer questions quickly and accurately. In the age of artificial intelligence, the most valuable worker is the one who can ask the right question.

Answers have become a commodity. With the rise of large language models and global information retrieval systems, the cost of knowing a fact has dropped substantially to near zero. You can summon the entire history of the Roman Senate or the code for a Python script in seconds. The ability to retain detailed information is no longer a competitive advantage in a world of apps and cheat sheets. The advantages now belong to the curious mind unafraid of rabbit holes.

This chapter argues that curiosity is not merely a whimsical trait of the reflective mood but a hard economic asset. It is the only cognitive function that cannot easily be automated. The machine can explore a database, but it cannot explore the unknown with intent.

The Inversion of Value

For the last century, our education and professional systems have rewarded specialization. We were taught to drill down, to narrow our focus, to become the indisputable expert in a shrinking domain. This was the strategy of the I-shaped individual (skills in one area). Think of the old joke about a PhD; the ability to know more and more about less and less.

However, the landscape has inverted. As David Epstein (2019) argues in *Range*, the modern world is wicked. It is characterized by ill-defined problems and missing data. In a wicked environment, the specialist is often at a disadvantage because they attempt to force new problems into old frameworks. The generalist, the amateur, and the curious operator thrive because they can draw analogies from disparate fields.

When you spend your sixteen hours reading obscure history, learning the mechanics of a watch, or studying the mycelial networks of fungi, you are not wasting time. You are building an internal database of analogies. When a crisis hits your specific industry, you will not look for the solution within the industry, because you know the industry's playbook is broken. You will look to the fungi, or the watch, or the history book. This cross-pollination is the engine of innovation (Johansson, 2004).

The Rabbit Hole as Strategy

We are often told to stay focused, to avoid going down the rabbit hole. This advice is a relic of the efficiency economy. In our new economy, the rabbit hole is the strategy.

The rabbit hole is simply a derogatory term for associative learning. It is the process of following a thread of interest wherever it leads, regardless of its immediate utility. Going to the moon spun off a plethora of new creations, Tang notwithstanding. You start by reading about coffee beans, which leads to the history of the Dutch East India Company, which leads to the economics of trade winds, which leads to naval architecture.

This nonlinear path creates a unique fingerprint of knowledge. No two souls go down the same rabbit holes. Therefore, your intellectual stack becomes difficult to replicate. If you only learn what is on the syllabus or if you only learn what is in the industry trade journal, you are running the same internal software as your competitors. When you follow your curiosity, you write your own code.

The Curiosity Coefficient

The most dangerous participants in any group or organization are the ones who have stopped asking why. This is a sign of intellectual calcification. It occurs when the mental model of a tribe or organization harden and refuse to accommodate new data or insights. See Kodak or Blockbuster or Radio Shack.

To combat this, cultivate a high curiosity coefficient. This is the willingness, indeed eagerness, to look foolish in the service of learning. It is the refusal to accept that is just how we do it as a valid explanation.

Ian Leslie (2014) distinguishes between diversive curiosity (the desire for novelty, like checking the entertainment news or reality TV) and epistemic curiosity (the desire to understand how things work). Diversive curiosity is the dopamine hit of the doomscroll or 24 hour news cycle. Epistemic curiosity is the sustained effort to resolve a gap in knowledge. The concentration core we built in Chapter Six is the tool that converts diversive curiosity into epistemic curiosity. It allows you to stay with the question long enough to find a genuine answer.

The Future-Proof Mindset

We cannot predict the specific skills that will be required ten years from now (assuming any skills will be required). A coding language learned today may be obsolete by next Tuesday. Therefore, just-in-case learning, learning a specific skill because it might be useful, is an increasingly risky bet.

The better bet is just-in-time curiosity. This is the confidence that, whatever the problem, you have the research skills, and the attention span, and the cognitive flexibility, to learn the solution in real time. You do not need to memorize answers, you do need to know how to find them.

Your unallocated equity is the training ground for this mindset. When you struggle to learn a new piece of music or understand a complex scientific paper on a Saturday morning, you are not just learning the content. You are cross-training your brain to be unafraid of the unknown. You are proving to yourself and the world that you are an adaptive organism.

To demonstrate that curiosity as a competitive edge is not just a romantic ideal but a driver of massive economic value, we examine three cases where a specific, useless rabbit hole led directly to market dominance.

Steve Jobs, The Calligraphy Class

The most famous example of the curiosity coefficient is Steve Jobs's decision to audit, for zero college credit, a calligraphy class at Reed College *after* he had dropped out. At the time, this curiosity had zero practical application. Computers were command line machines with green text. Jobs was simply fascinated by the history of serif and sans serif typefaces.

Ten years later, when designing the Macintosh, this useless data resurfaced. Jobs integrated his knowledge of typography into the operating system, creating what Jobs later described as the first computer with beautiful fonts. This aesthetic differentiation became Apple's primary competitive castle moat, separating it from utilitarian IBM and Microsoft. It established a market capitalization that would eventually be measured in the trillions. As Jobs famously noted in his 2005 Stanford University commencement address; "You can't connect the dots looking forward, you can only connect them looking backwards. So you have to trust that the dots will somehow connect in your future. You have to trust in something, your gut, destiny, life, karma, whatever. This approach has never let me down, and it has made all the difference in my life" (Jobs, 2005).

Dasia Taylor and the Beet Juice Suture

At age 17, Dasia Taylor became obsessed with a problem of fairness. Smart sutures (stitches that use electronic sensors to detect infection) were too expensive for developing nations. Her rabbit hole was not high tech engineering, but the chemistry of the grocery produce aisle. She spent months testing various vegetable juices, fascinated by the useless fact that certain root vegetables change color based on pH levels.

While her peers were coding apps, Taylor was in her kitchen boiling beets. She discovered that beet juice turns from bright red to dark purple at a pH of 9, the exact acidity level of an infected human wound. She realized she didn't need microchips. She just needed to dye cotton thread with beet juice.

Her invention, color changing sutures, provides a visual early warning system for surgical site infections that cause significant mortality and morbidity in low income countries. These sutures cost pennies on the dollar compared to electronic alternatives. This low tech curiosity gained her global recognition as a finalist at the Regeneron Science Talent Search and led to plans to patent a device that empowers patients to diagnose their own infections simply by looking at their skin (Machemer, 2021).

Hedy Lamarr and the Player Piano

While the world knew her as the most beautiful woman in Hollywood, Hedy Lamarr spent her nights in a useless hobby; tinkering with the mechanics of avant-garde music. She became fascinated by the player piano, specifically, the way a perforated paper roll could control the keys.

During World War II, she realized that radio-controlled torpedoes were easy to jam because they stayed on one frequency. Drawing directly from her musical hobby, she sketched a system where the transmitter and receiver would hop between frequencies like a player piano jumping between notes.

The US Navy bureaucracy (Peter Principle?) originally rejected the idea as unfeasible, telling her to sell war bonds instead. However, her patent for frequency hopping spread spectrum invention eventually became a backbone of secure digital communications. Today, it remains the underlying technology for Bluetooth and was a direct precursor to early Wi-Fi standards. That multi-trillion-dollar ecosystem derived from a movie star's curiosity about a piano (Rhodes, 2011).

James Dyson, The Sawmill Anomaly

In the late 1970s, James Dyson was frustrated by his vacuum cleaner's loss of suction. Serendipity occurred while visiting a local sawmill. His curiosity was piqued by a giant industrial cyclone on the roof. It used centrifugal force to separate sawdust from the air.

Instead of ignoring this detail as sawmill trivia, Dyson engaged in cross domain analogy (pairing the architecture of two unrelated systems). He asked a question no one ever asked in the vacuum industry before. Could this large capacity industrial technology be scaled down to fit in a home appliance? This spark of curiosity led to over 5,000 prototypes and the invention of the bagless vacuum. He imported an idea from a completely unrelated industry (timber). Dyson created a multi-billion dollar company and revolutionized a stagnant market (Dyson, 1997).

Katalin Karikó and the Fragile Molecule

For decades, the Hungarian biochemist Katalin Karikó was obsessed with messenger RNA (mRNA), a molecule considered by the scientific community to be too fragile and unstable to ever be used as a drug. Her obsession was viewed as a career dead-end. She was demoted, her pay was cut by Peter Principled leadership. She was mocked for pursuing useless science while her peers focused on DNA.

Karikó refused to abandon this rabbit hole. She spent years figuring out how to shield the mRNA so the body wouldn't destroy it. This obscure, technical tweak, modifying a single nucleoside, turned the fragile molecule into a programmable instruction manual for the human body.

When the COVID-19 pandemic hit, her useless discovery became the foundational science for the ground breaking

BioNTech/Pfizer and Moderna vaccines. This not only saved millions of lives but created a massive new biotechnology sector, shifting the entire pharmaceutical paradigm toward mRNA therapies (Karikó, 2023). Without Karikó, Operation Warp Speed would still be chugging along like a snail. New morgues would have had to be opened.

Kevin Systrom, The Holga Lens

Before founding Instagram, Kevin Systrom was a marketer who spent his evenings learning to code (a rabbit hole in itself). But his real competitive edge derived from his curiosity about fine art photography. Specifically, his study of the Holga camera during a semester in Florence. The Holga was known for its blurry, artistic, square format images.

When his partner pointed out that their early app's photos looked horrible compared to professional images, Systrom didn't look for a better compression algorithm (the specialist's answer). He looked to his useless knowledge of art history. He realized that the flaws of the Holga could be a feature and a benefit. He created the X-Pro II filter to mimic that artistic style. This feature, born of artistic curiosity, not engineering efficiency, was the catalyst that allowed Instagram to dominate the social media landscape and eventually sell for $1 billion (Frier, 2020).

Sara Blakely and the Scissors

Sara Blakely was not an engineer; she was a fax machine salesperson who was irrationally annoyed by the problem of visible panty lines under white pants. Her curiosity was not high tech, but tactile. She spent hours cutting the feet off pantyhose trying to find a way to smooth the figure without the discomfort of a girdle.

Every manufacturer she called rejected her, and patent lawyers thought she was a prank caller. She wrote her own patent using a textbook from Barnes & Noble. She relentlessly pursued the footless concept that the hosiery industry deemed unnecessary.

By solving a problem that male-dominated industry leaders didn't know existed, Blakely created Spanx. This did not just launch a product, it invented the modern shapewear category. It lead to a billion-dollar valuation and landing Blakely on the cover of Forbes as the youngest self-made female billionaire of her time (O'Connor, 2012).

The Ultimate Hedge

Curiosity is the ultimate hedge against obsolescence. Machines are built to answer. Humans are built to wonder, to ponder. As long as one remains interested in the wonders of the world, as long as you refuse to let your eyes glaze over, you will have a place in it.

The curious mind is an ever nimble mind. It does not cling to the sinking ship of an outdated career. It looks at the horizon, observes that change coming, and asks, I wonder how that works?

Figure 1

Chapter Eleven

The Hazards of the Prig

WE have reached a rather delicate but essential point in our reconstruction of the day. If you have followed the protocols of the previous ten chapters, you are now waking with greater purpose, experimenting with deep work, and utilizing your sixteen hours with goals tuned with the precision of a Swiss watch. You are reading denser than usual subjects while your neighbors watch reality television. You are forming new curiosity driven skills while your colleagues doomscroll or scurry to the next ping.

You are, objectively, using your time better than the majority of the population. Even inhabiting the 75th percentile with higher aspirations is a noteworthy achievement (worthy of another pat on the head by a human being).

And this is where the trap snaps shut. The moment you begin to feel a sense of moral superiority over the time-wasters around you, you have contracted the disease of the *prig*. In his 1910 text, Arnold Bennett warned explicitly of this danger, noting that a prig is "a pert fellow who gives himself airs of superior wisdom" (Bennett, 1910, Chapter XII).

Let Webster enlighten you. Prig; "one who offends or irritates by observance of proprieties (as of speech or manners) in a pointed manner or to an obnoxious degree"(merriam-webster, n.d.).

If your quest for a better life makes you unbearable to live with, you have not succeeded. You have merely swapped indifference for obnoxious arrogance (a difficult point to make with humility).

The Mathematics of Ego

The prig is born from a miscalculation of value. The prig believes that because they have read a Plato Platonic with marginalia before breakfast, they are empowered to look down upon the person who slept in. This is a critical error.

The discipline you are building is a tool for your own liberation and edification, not a weapon for social judgment. When you use your newly cultivated insights to judge others, you are violating the reflective mood. I would suggest you use the reflective mood as your first commandment. Do not turn your gaze outward to critique the world, rather focus all your energy inward to build the self.

More important, the prig is fragile. Their discipline relies on the contrast with others. They need to see others failing to feel successful. It is a purely selfish act to place oneself on a pedestal. A truly reflective mood should have little difficulty unearthing a number of personal flaws that need work. A truly sovereign mind does not care what the rest of the world is doing, because the pursuit of continuous improvement is internal.

The Domestic Insurgency

The most dangerous habitat for the prig is the home. There is a tendency, once one has seen the light of the sixteen hours, to become an evangelist to one's spouse or family. Bad move grasshopper! You decide that we should all stop watching television. You decide that we should all learn French at dinner.

This is a counterproductive strategy. It leads inevitably to domestic friction. Your partner did not sign a contract for your self-improvement regimen. They have their own agency, their own fatigue, and their own needs. If you do not see that reread the discussion on empathy in Chapter 8. The discussion of utilitarianism vs empathy belongs elsewhere, not in your immediate circle.

The rule here is absolute. Your schedule is a private monarchy. You have no authority or jurisdiction over the time of others. If you wish to read while your spouse watches a sitcom, you must do so without a heavy sigh or a roll of the eyes. You must do it with total neutrality. If your example inspires them, it is excellent. If it does not, that is also excellent. You are doing this for you. At best, leave a copy of this book in a convenient place and let nature take its course.

The Protocol of Silence

To inoculate yourself against priggishness, you must adopt the protocol of silence. This is the Fight Club rule of life architecture. The first rule of the sixteen hours is that you do not talk about the sixteen hours. If you feel overly wounded by this protocol visit a support group for testicular cancer victims to see what real suffering is like (Palahniuk, 1996).

Do not post your morning routine on social media. Do not announce at the water cooler that you have given up the news. Do not tell your friends and family how productive your weekend was unless they explicitly ask.

When you perform your discipline in public, you are seeking validation. You are asking for a pat on the head from a human being. This externalizes your motivation while internalization is the entire point of the affair. If the likes do not come, will the discipline fade? Reboot time. True intellectual sovereignty is practiced in the dark. It is the work you do when no one is watching, and no one will ever know.

The following examples illustrate the social and psychological costs of becoming the prig, where the pursuit of self-optimization crosses the line into isolation, domestic friction, or disconnection from shared humanity.

Bryan Johnson, The Solitude of the Blueprint

Bryan Johnson, the tech centimillionaire, is the current pinnacle of the 21st century optimized life. He has been reported to spend approximately $2 million annually on his blueprint protocol (Vance, 2023). His publicly reported regimen has involved taking more than 100 supplements daily and eating his last meal by 11:00 AM (intermittent fasting) (Vance, 2023). While he has reported achieving the biological markers of an 18-year-old in specific metrics such as lung capacity and inflammation (he is 48 years old in 2026) (Vance, 2023), he has inadvertently become a case study in the mathematics of ego.

Johnson openly admits that his lifestyle makes traditional social connection impossible. He cannot share a dinner with friends (he doesn't eat dinner), he cannot have the occasional casual drink (alcohol is poison), and he cannot stay out late. He has effectively optimized himself out of the human tribe. In interviews, he describes a calculated data led existence that, while impressive, strikes many observers as a lonely, sterile prison of one's own making (Vance, 2023). This demonstrates the hazard. When the concentration core becomes absolute, it builds a wall that keeps the world out.

A.J. Jacobs, The Domestic Insurgency

Author A.J. Jacobs spent two years striving to be the healthiest man alive for his book *Drop Dead Healthy*. He adopted every piece of advice he could find. He experimented with chewdaism (chewing food up to 50 times per bite), wore a helmet while walking, and installed a treadmill desk. He became the ultimate self-improvement project.

Gemini creation

However, the real lesson of his experiment was the domestic insurgency. His wife, Julie, did not view his transformation as inspiring; she viewed it as utterly insufferable. She noted throughout the experiment that his health obsession was becoming a constant point of annoyance. Jacobs realized that his obsession with personal optimization was imposing a harsh tax on his marriage. His behavior dialed in priggishness to an art form. He brought his own health food to parties, shushed people to protect his ears, and lectured on posture. He transformed to the consummate prig to the very person who mattered most. He had to dial back his protocols to save his relationships, proving that a schedule that alienates your family and friends is a failed schedule (Jacobs, 2012).

Paul Miller, The Offline Superiority Complex

In 2012, technology writer Paul Miller disconnected from the internet in totality for one year. Initially, he fell into the trap of the prig. He felt a sense of smug moral superiority over his connected peers. He believed he was living a deeper, more authentic life because he was reading Greek literature while his friends were tweeting (X-ing?).

However, six months later, the prig demeanor crumbled. Miller found that his disconnection did not really make him a philosopher king. It made him lonely. He lost touch with the cultural nuances of his friends. The inside jokes went over his head. He lost track of the memes and the shared digital empathy. By the end of his experiment he had concluded that the internet, warts and all, was where the humans were. To reject it entirely was to reject society. He realized that the goal is not to exit the room (and brag about it), but to learn how to live in the room without being consumed by the attention economy (Miller, 2013).

The Antidote of Ignorance

The surest cure for priggishness is a hefty dose of perspective. You may have read six books this month. There are more than thirty-two million cataloged books in the Library of Congress (Library of Congress, n.d.). You may have learned the essentials of Python. There are teenagers who can code circles around you in their sleep.

The more you learn, the more you should realize the *terrifying scope of your own ignorance.* The human who truly engages in deep reading (Chapter Eight) or synthetic work (Chapter Nine) does not become an arrogant prig. They become humbler. They see the Everest of the mountain they have yet to climb.

The prig stands on a molehill and thinks they have conquered the world. The anti-prig stands on the mountain and realizes they are less than a speck of dust within the universe

Be the philosopher. Do the work. Keep it quiet. And be kind to those who are still sleeping. You were one of them only yesterday.

Do Not Be A Prig!

Chapter Twelve

The 24 Hour Ecosystem

WE have nearly reached the end of our architectural and hopefully inspirational project. We have disassembled the mechanisms of the day. We lubed the patinaed gears of the concentration core. We buffed up the springs of the reflective mood. We replaced the hour hand of the sixteen hours.

Now, we must reassemble the refurbished watch.

The danger at this stage is to view these twelve chapters (plus one) as a checklist of rigid obligations. If you view the morning startup, the deep reading sprint, and the evening review as separate, heavy stones that you must trudge along with every day, you will collapse under their weight within days.

We must shift our metaphor one last time. You are not constructing a fine-tuned machine. You *are* cultivating a personal environment. A machine is brittle; if one part breaks, the whole stops. An environment is resilient. If one element fluctuates, the others adapt to maintain balance. This chapter outlines how to maneuver these practices into a 24 hour ecosystem that can survive the chaos of real life.

The Feedback Loops of Sanity

In a healthy ecosystem, the waste products of one process can become the fuel for another. In your updated and possibly rebooted life architecture, the energy expended in one phase could and/or should recharge another.

Consider the relationship between the concentration core (Chapter Six) and the reflective mood (Chapter Seven). They are not separate tasks. They are a polarity. The passion or intensity of your deep work mandates the need for the calm of the reflective mood. The peace of the reflective mood regenerates the dopamine required for the next phase of deep work.

If you find yourself torn or exhausted, do not ask, what is wrong with my willpower? Your willpower is just fine and always has been. You need to ask, where is the leak in my environmental ecosystem? You are likely drawing down on your reserves without replenishing the system's essential stocks or buffers (Meadows, 2008).

The Principle of Plasticity

Arnold Bennett (1910) warned explicitly against the danger of becoming a slave to the programme. He understood that the rigid adherent, having set a color coded detailed schedule for self-improvement can and most certainly will, experience interruptions. If one then suffers a breakdown of the environment then the purpose of the system has been defeated.

Your ecosystem should, indeed must, possess plasticity. There will be days when the threshold protocol fails. There will be weeks when the deep reading is replaced by the flu or a family crisis.

The novice quits when the schedule breaks. The architect understands that the schedule is just a map, not the terrain to be negotiated. When the ecosystem is disturbed, you do not abandon it. Take a deep breath (maybe more) and engage in re-centering. You gently steer the ship back to the course the next mornings. You do not overcompensate by reading *War and Peace* for eight hours to make up for the lost time. Just resume the rhythm. The old cliché works. It is a marathon, not a sprint.

The Keystone Species

In ecology, a keystone species is an organism that holds the entire system together. If you remove it, the framework crumbles. In your new 24 hour environment, you must identify your keystone habit.

For some, it is the morning startup (Chapter Four). If the first thirty minutes of the day are secured, if the coffee is drunk in silence, the phone is ignored, and the plan is mentally reviewed, the rest of the day tends to follow the reflective mood. That can occur even if chaos ensues at the office.

Others anchor the evening review (Chapter Seven). No matter how badly the day went (and they can go badly), the act of closing the books and clearing the cognitive residue facilitates tomorrow's sunshine rising on a fresh inheritance.

Find your keystone. Protect it with your sanity and your life. Personally, I will never rank my day in a spreadsheet, or start the day without glancing at the weather on my now evil cell phone. But, by absorbing the wisdom of Bennett I can ignore news feeds and pings like a champ now. If you can only do one thing from this book on a bad day, do that one thing. Strive to do more on the good days.

The Compound Interest of Being

Why, why, why go to this trouble? Why, why, why fight the algorithm? Why, why, why embrace the friction?

The answer is in the compound interest of being. When you invest twenty minutes a day in the internal archive, or thirty minutes in synthetic work, the returns are not linear. They are exponential over time, just like compound interest.

In the first months or year, you are a person who knows a little more about history than your neighbors. In year four you are a human with a fundamentally different operating system and quite likely a higher IQ should you test for that. You have a deeper reservoir of wisdom, a sturdier fortress of focus, and a deeper depth, a calmer pool, of character that the reactive world cannot penetrate as deeply. You have become, in the truest sense, an individual (Clear, 2018).

History provides us with three avatars who successfully lived the exact architecture Bennett proposed. They maintained a primary career while building immense stature through adherence to improving the margins of the day.

Anthony Trollope, The Morning Protocol

The Protocol: Bennett argued that the time before the workday is the only unsoiled currency of the day and must be spent on the evolution of the immortal soul.

The Embodiment: Anthony Trollope is the gold standard for this. While working a demanding full time job as a surveyor for the British post office, Trollope rigorously adhered to a morning routine. He paid an attendant to wake him at 5:30 AM every day with coffee. He then wrote for exactly three hours, monitoring his watch to produce two hundred fifteen words every fifteen minutes, before leaving for his professional duties.

The Result: By reclaiming the morning, he produced an immense body of work, dozens of well received novels, while rising to the top of the postal service. He proved Bennett's thesis. The practitioner workday does not have to consume the creative soul (Trollope, 1883).

Wallace Stevens, The Commute Protocol

The Protocol: Bennett specifically advised using transit to work not for reading newspapers, but for a concentration of the mind, forcing the brain to hold a single complex thought or rhythm to build mental muscle.

The Embodiment: Wallace Stevens, a lawyer and high-powered insurance executive (Vice President of the Hartford Accident and Indemnity Company), used his daily walk to the office to compose poetry. He did not write on paper while walking. Instead, he used the rhythm of his stride to concentrate the lines in his head, scribing them only when he reached the office. From the rhythm of his steps, "Reality is a cliché from which we escape by metaphor" (Vintage, 2011, p. 260).

The Result: He became one of America's greatest modernist poets while maintaining a 40-year corporate career, perfectly executing Bennett's commute protocol (Brazeau, 1983).

T.S. Eliot, The 90-Minute Evening Protocol)

The Protocol: Bennett suggested setting aside three evenings a week for 90 minutes of serious reading or creative synthesis, treating this time as sacred as a business appointment.

The Embodiment: During his eight years at Lloyds Bank, T.S. Eliot did not view his banking work as his identity. He compartmentalized his professional labor (handling foreign accounts) and used his evenings and weekends for synthetic work. He reportedly drafted parts of *The Waste Land* during these appointed blocks of time, often while recovering from the exhaustion of the bank (recharging in the process).

The Result: He proved that one can be a harassed practitioner and still produce one of the most significant poems of the 20th century by guarding the day within a day (Gordon, 1999).

The Final Ownership

We return, in due course, to the premise with which we began. You wake up every morning with the 24 hours of unmanufactured tissue of the universe of your life. It is yours. It is the most precious of possessions.

No one can take it from you. The boss can take your labor, the bank can take your house, but the daily miracle is untaxable and unstealable. It is delivered to you, fresh and unspoiled, every dawn.

You now have the tools to inhabit it. You have the architecture to structure it. You have the philosophy to understand it.

The door is open. The house is ready. Live.

Chapter Thirteen

The Infinite Game

WE began this book with a metaphor of accounting. We spoke of 24 hours as a currency, unallocated equity, and of the daily miracle. But every accountant knows that a balance sheet is only a snapshot in time. Equity = Assets - Liabilities. The truest measure of wealth is not what you hold in your hand today, but what you can create and sustain across a lifetime. Start by not clinging to regrets. You know not what ripples you have created in the fabric of time.

Ripples

I have nearly zero memories from first grade (aside from a girl named Marilyn with long brown hair, one row over and two seats up). But somehow, burnt into my memory was an impassioned speech by a student teacher interning for a few days. She described slavery and the history of it with heartfelt emotion. The word whips entered into the homily. I can still picture the room and where she stood. Then, our regular first grade teacher interrupted her. She took a more sympathetic take on slave ownership stressing the "culture" of the South. Setting the regular teachers apologist views aside, that speech planted a seed of the woke mentality that I take pride in embracing and owning to this day. The anti-woke crowd plays the ultimate finite game. Time and time again history has vilified those kinds of prigs. Empathy and education are not an evil to be supplanted by the greedy tribal and narrow mind. Point being, I cannot remember that interns name but I will never forget her interrupted speech and the passion she put in it. Ripples. They can spread for a very long time.

Conversely

Ripples. The worst among us have produced unintended consequences of great good.

The willful and calculated directive by Max Blanck and Isaac Harris to physically chain the exit doors of the Triangle Shirtwaist Factory in 1911 stands as a verifiable tragedy engineered solely by adherence to absolute industrial avarice (Von Drehle, 2003).

Maintaining the ninth floor of the Asch Building as an impregnable fortress for garment manufacturing, these businessmen prioritized the eradication of petty theft over elementary human preservation. They deliberately severed the singular route of internal egress (Argersinger, 2009). This premeditated act transformed a routine production floor into a death trap enclosure, ensuring that a scrap bin fire on March 25, 1911, escalated into a massacre that eliminated 146 lives (Greenwald, 2005). Their conscious, willful choice to lock the workforce inside constituted a sociopathic humanity tax levied against immigrant laborers to secure maximal inventory retention (Von Drehle, 2003).

The immediate consequence of this butchery triggered an aggressive transfer of political power from localized Tammany Hall ward bosses to institutional architects like Frances Perkins (Downey, 2009). Perkins utilized the visceral horror of the falling bodies to shatter the established architecture of laissez-faire capitalism, forcing a lethargic state apparatus to acknowledge that unchecked industrial autonomy equated to state sanctioned homicide (Argersinger, 2009). Regulation sometimes requires blood.

The owners' precision cruelty catalyzed the creation of the Factory Investigating Commission, which subsequently dismantled the era's pure capitalistic priority of voluntary corporate compliance by enforcing thirty-six stringent industrial labor laws within three years (Greenwald, 2005). This immediate production of statutory mandates codified mandatory fire suppression systems, emergency egress protocols, and unannounced inspections. The fire inadvertently engineered the legal and physical framework that currently preserves millions of workers globally (Von Drehle, 2003). Cruelty codified survival.

Move Finite to Infinite

Most folks treat time as a finite game. As James Carse (1986) defined it, a finite game is played for the purpose of winning. We attempt to win the workday by clearing the inbox while the shot clock runs down. We try to win the weekend by jamming in activities. Practitioners treat the 24 hours as a contact sport to be conquered.

But life is an infinite game (until it isn't). The purpose of an infinite game is not to win, but to continue to play the game. You do not win a marriage. You do not win a friendship. And you most certainly do not win a life. You only get to keep playing until entropy redefines the game. This final chapter is a call to shift your perspective from the finite metrics of productivity to the intangible, infinite, distant horizon of meaning.

Beyond the Professional Avatar

For decades, you have likely curated a professional avatar. This is the version of you that exists on LinkedIn, in a company directory, and in the minds of your friends and colleagues. It is likely efficient, competent, and perhaps highly successful. But it is not you.

The tragic error of the modern age is the conflation of the resume virtues with the eulogy virtues (Brooks, 2015). We spend our sixteen hours polishing the resume virtues. The skills that bring income and status. We typically neglect or subordinate the eulogy virtues. The ripples of character, the ripples of kindness, the ripples of curiosity. These are the virtues that spawn the ripples. Those ripples spread long and far after the reaper with a scythe has moved on. Who are you to predict the shore those ripples reach?

The concentration core and the reflective mood are not designed to make your avatar bright and shiny. They are designed to help you survive, in some way and fashion, the *inevitable* death of that avatar. One day, the title will be stripped away. The email account will be deactivated. What ripples will be left spreading in the sound of silence? If you have invested your unallocated equity wisely, the person left standing will be a scholar, a creator, a friend, and a whole human being. If you have not, the person left standing might be a ghost.

Yvon Chouinard, The Exit from Capitalism

Yvon Chouinard, the founder of Patagonia, spent fifty years wrestling with the finite game of business. Five decades. In the standard finite game the goal is to maximize shareholder value and prepare for the big payday (selling the company or going public). Chouinard realized that playing by these rules would eventually destroy the mission of his company.

In September 2022, Chouinard made a move that baffled the financial world. Instead of selling the company for its valued $3 billion, he transferred 2% of the stock (all of the voting shares) to the Patagonia Purpose Trust and the remaining 98% (non-voting stock) to the Holdfast Collective, a non-profit dedicated to fighting climate change.

By doing this, Chouinard effectively snapped the finite game in half. He ensured that the company could never be sold. He ensured that the company could never be stripped for assets. He ensured that the company could never be diverted from its values by future shareholders. He removed his professional identity as a billionaire to help try and secure the infinite legacy of the planet. As he stated, instead of extracting value from nature and transforming it into wealth for investors, “we’ll use the wealth Patagonia creates to protect the source of all wealth” (Gelles, 2022). Ripples.

Jimmy Carter, The Architecture of the Afterlife

Jimmy Carter is the quintessential example of a man who lost the finite game but won the infinite one. In 1980, he endured a crushing defeat in the finite game of presidential politics. By the metrics and values of Washington, he was a failure. A one-term president rejected by the voters.

However, President Carter refused to let his professional identity, his avatar, define his existence. Unlike his peers who moved into the finite games of board seats and speaking fees, Carter entered the infinite game of humanitarian service. Through the Carter Center, decades were spent working to eradicate Guinea worm disease, monitor elections in fragile democracies, and build houses with Habitat for Humanity to fill a desperate need.

In 2002, he was awarded the Nobel Peace Prize, not for his presidency, but for his after-avatar, his infinite game. Carter demonstrated that a title is just a temporary role in a finite game. Character is the engine of the infinite game. He proved that one's greatest work often begins only after the professional avatar has been retired. Ripples.

Rose Marcario (Patagonia)

A COO/CEO focused on supply chains and operations.

Marcario took Chouinard's vision and operationalized it into radical action. Under her leadership, the company challenged the government's stance on public lands and proved that corporate operations can be a vehicle for activism, using the company's infrastructure to drive civic engagement rather than just profit. Ripples.

Danny Hillis and The 10,000 Year Clock

In an era defined by the nanosecond speed of the internet, computer scientist Danny Hillis proposed a project to force humanity to think in deep time. He recognized that our obsession with the now (the finite game of speed) blinded us to the long term consequences of our actions.

To counter this, Hillis and the Long Now Foundation conceived the 10,000 Year Clock. This clock is designed to tick once a year, with the century hand advancing once every one hundred years, and the cuckoo coming out on the millennium.

The clock has absolutely no commercial value. It fails to solve any immediate problem. It is a piece of synthetic work on a monumental scale. A rabbit hole of sorts. Its sole purpose is to serve as an icon for long term thinking. It encourages visitors to extend their mental horizon beyond their own lifetime (way beyond). By building a mechanism designed to outlast civilization as we know it (and potentially all of civilization), Hillis is playing the ultimate infinite game, creating a context for generations he will never meet (Brand, 1999). Ripples.

Satoru Iwata, The Nintendo Shield

In 2013, the video game giant Nintendo was facing a financial crisis. The launch of the Wii U console was a commercial failure, and profits had tanked. In the finite game of corporate management, the standard move is a restructuring. Restructuring is a polite (cruel?) euphemism for mass layoffs to balance the short-term books.

Satoru Iwata, the company's CEO, adamantly refused to play this game. Holding to the philosophy that fear of layoffs destroys the critical morale required for innovation, Iwata chose a different path. Instead of firing staff, Iwata took a voluntary 50% pay cut, and the board took reductions of 20-30% (Ashcraft, 2014). By protecting his workforce (the infinite asset) at the expense of executive compensation (the finite metric), Iwata preserved the company's creative culture. Three years later, that same team released the Nintendo Switch, one of the most successful consoles in history to date. Ripples.

Lisa Su (AMD), A corporate administrator managing a failing semiconductor company.

An engineer who refused to stop being an engineer.

As perfect reflection of Satoru Iwata, Su saved AMD not by financial engineering (shadow work), but by returning the company to innovation. She is often credited with refocusing the company's culture on high performance computing and engineering excellence rather than short-term stock manipulation, proving that stature comes from understanding the product. Ripples.

Reshma Saujani (Girls Who Code), A failed congressional candidate and lawyer.

A movement builder for *Brave, Not Perfect.*

After losing her New York congressional race by 62 percentage points, Saujani realized she had spent her life trying to be perfect (the ultimate Avatar). She pivoted to found Girls Who Code, teaching young women to embrace the bugs in their code as well as their lives. She encouraged them to take risks. She defines her stature not by the election she lost, but by the cultural shift she ignited (Saujani, 2019). Ripples.

Bob Chapman, The Shared Sacrifice

During the 2008 financial crisis, the manufacturing firm Barry-Wehmiller lost 30% of its orders overnight. The board's finite game calculation was simple. They needed to save $10 million, which mathematically equaled laying off a significant portion of the workforce.

CEO Bob Chapman refused. He operated on the infinite principle that we measure success by the way we touch the lives of people. He proposed an alternative. His proposal and action was a mandatory furlough program where every employee, from the CEO to the receptionist, had to take four weeks of unpaid leave. He framed it with the slogan, "It is better that we should all suffer a little, than that any of us should suffer a lot" (Sinek, 2014).

The result was unexpected. Morale improved. Employees swapped weeks to help those who could not afford the loss. The company not only survived the recession without a single layoff but emerged with a level of trust and loyalty that drove record profitability in the following years. No Peter Principle here, just ripples.

Jimmy Wales, The Billion-Dollar Refusal

In the early 2000s, Jimmy Wales found himself sitting on one of the most valuable digital assets in the world, Wikipedia. As traffic exploded and venture capitalists and advertisers circled, he was offered the finite game reward. An IPO that would have likely made Wales a billionaire.

Wales recognized that introducing the profit motive (the finite game) would forever corrupt the neutrality of the encyclopedia (the infinite game). An algorithm that was optimized for ad revenue *always* prioritizes engagement over truth (the attention economy). To protect the integrity of human knowledge, he effectively burned the ships, structuring Wikipedia as a non-profit dependent on donations (chip in plz). He traded a personal fortune for the infinite preservation of a public utility. That action ensured that the platform would remain a servant to the user rather than a servant to the shareholder (Cooke, 2020). Ripples.

The Legacy of Attention

We often think of legacy as something we leave behind, money, buildings, or children. But your true legacy is where you chose to place your attention. The ripples you created. Simone Weil famously wrote that the capacity to give one's attention to a sufferer is a very rare and difficult thing; it is almost a miracle (Weil, 1950/2009).

When you choose to spend an hour listening to a friend instead of scrolling your phone, you are building a legacy and creating a ripple. When you choose to read a difficult book instead of zoning out to a screen, you are preserving the legacy of human thought and receiving a ripple. The 24 hour ecosystem you have embraced by reading this far is a mechanism for directing this generosity.

Every time you engage in deep reading, you vote for a world that values depth. Every time you spurn the algorithm, you are voting for human autonomy. Your life is a series of votes and not just at the ballot box. Cast them with intention.

The Steward of the Fortune

You are now the steward of a great fortune. You wake up every morning with the same 24 hours that were given to Michelangelo, to Einstein, and to whomever the person is that you admire most. There is no poverty of time; there is only a poverty of attention.

Do not let this fortune slip through your fingers in an algorithmic stream of digital distractions. Do not let it be stolen by the attention economy filled with the demands of others (that have zero interest in your wellbeing). Do not let it be stolen by the friction of a disorganized mind. Guard it. Cultivate it. Invest it in the things that matter.

A Final Call to Action

The book is nearly closed. Bennett's Theory is before you. The clock on the wall is ticking, ticking, ticking, indifferent to your plans. You have new tools. You understand the architecture. You know the cost of the drift and the value of the focus.

The next hour is arriving by the second. It is a fresh, unblemished page. It asks nothing of you, yet it demands everything.

Do not just manage your time. Honor it.

Step out of the reactive current and onto the solid ground of your own existence. The infinite game is waiting for your ripples.

APPENDIX

Glossary of Terms

Active detachment: The deliberate effort to disconnect the mind from shadow work (low value work) within the practitioner sphere. Unlike passive rest, active detachment involves engaging in synthetic work (personal enrichment activities like bread-making, study or DYI labor etc.). That requires a different form of cognitive load allowing the concentration core to recover from the day's technical friction (Sonnentag & Fritz, 2015).

Assembly and e-Discovery: The process in legal and corporate sectors for identifying, collecting, and producing electronically stored information in response to a request for production in a lawsuit or an investigation. This process imposes a technical friction tax, because practitioners must navigate vast quantities of digital data and complex software interfaces rather than focusing on core legal strategy (Susskind, 2019).

Automated Document Assembly: The use of software to create boilerplate legal and/or administrative documents through predefined templates and data inputs. Intended to save time but often forces practitioners into shadow work, where they must troubleshoot formatting errors and metadata ghosts created by the automation (Remus & Levy, 2017).

Cognitive autonomy: The fundamental right and ability of an individual to exercise self-determination over their own mental processes. The cognitive autonomy is a shield that allows a practitioner to resist algorithmic manipulation and maintain their own mental supply chain (The Denny Center, 2025).

Cognitive endurance: The developed capacity to sustain a single, complex line of inquiry or high-value task over an extended duration without succumbing to context switching or the impulse for digital distraction. This is a primary metric of the sovereign mind and is the direct antithesis of the fragmentation caused by modern interface design (Newport, 2016).

Cognitive Priming: The act of preparing the brain's neural pathways for a specific high value task. Unlike the process of getting to work, cognitive priming involves low-friction rituals, such as reading a single page of a difficult text or reviewing a handwritten goal. That signals the concentration core to transition from a state of rest to a state of direct production. This forestalls the mental stutter that occurs when attempting to enter a sacred appointment without a psychological on-ramp (Duhigg, 2012).

Deep Integration: The process of moving a high level insight or a newly acquired skill from the short-term storage of the digital interface used into the long-term architecture of the practitioner's life. In the context of the modern practitioner, deep integration is the antidote to shadow work. It ensures that information is not merely consumed but is fundamentally woven into one's useable memory and personal character (Newport, 2016).

Digital commute: The intentional mental transition period between waking and the start of labor (direct production). While a physical commute involves travel, the digital commute is an internal journey where the practitioner avoids the update trap and social noise to arrive at their desk with their intellectual sovereignty intact. Remote work requires ritual on/off switching. Failing to manage the digital commute often leads to digital submission before the workday begins (Perlow, 2012).

Digital Fast: A period of total abstinence from digital interfaces and notifications. This is a defensive maneuver used to reset the brain's dopamine receptors and reclaim cognitive endurance. It clears the digital cobwebs that obscure long-term strategic vision (Newport, 2016).

Digital Submission: A state of existence where the practitioner's attention is influenced by the algorithm. In this state, the individual is no longer a sovereign practitioner but a servant of the interface, reacting to notifications and performing shadow work under the illusion of connectivity. digital submission is the ultimate loss of cognitive autonomy (Lambert, 2015).

Direct production: The phase of labor where value is actually created, such as drafting documents, coding a core algorithm, or processing physical materials. (Lambert, 2015).

Emotional Autonomy: The preservation of one's internal emotional state against the external outrage loops of the digital world. It involves the disciplined refusal to allow the 24-hour news cycle or social media feedback loops to dictate one's mood or sense of security, acting as a defense against headline stress disorder (Stosny, 2017).

Frictionless waste: The phenomenon where removing the physical effort or cost of a task leads to an explosion in its volume and a devaluation of the resource involved. In the digital office, the zero cost of CC: All or Instant Message results in a surplus of low value communication that drowns out meaningful work (Sellen & Harper, 2002).

Gap: The specific, often fleeting moment of silence that exists between an external event (the Stimulus) and your personal action (the Response). Think ping or chat message. This space is the only territory where intellectual sovereignty truly lives. If a practitioner ignores the gap (immediately responding), they become a biological robot, instinctively snapping back at every notification or insult. However, by widening this gap, one gains the power to choose a response that aligns with their life architecture. (Frankl, 1946/2006).

Hard Temporal Firewall: A non-negotiable, time-based boundary that separates professional labor from personal life. It is the practice of locking the door to the digital office at a specific hour, preventing the time debt of the workplace from leaking into the hours reserved for family, rest, or personal synthetic work (Perlow, 2012).

Headline Stress Disorder: A state of chronic high arousal and anxiety triggered not by immediate physical danger, but by the relentless, algorithmic delivery of global crises into the private sphere. It is characterized by a feeling of helplessness and permanent vigilance caused by the 24 hour news cycle (Stosny, 2017).

Inbox zero: A rigorous email management philosophy that aims to keep the inbox empty at all times. This often functions as a deceptive form of shadow work. While originally intended to reduce stress, it frequently forces the practitioner to prioritize the administrative clearing of frictionless waste over high-value cognitive output, effectively turning the practitioner into a servant of the interface (Newport, 2021).

Input Independence: The active, conscious curation of one's information diet. It is the practice of selecting primary sources and nutritionally dense data (such as books or peer-reviewed research) while rejecting the algorithmic drift of "for you" feeds, news cycles and automated recommendations (Iosub, 2025).

Intellectual sovereignty: The overarching state of being an independent generator of ideas rather than a passive processor of external feeds. It is the ultimate goal of the practitioner, achieved by mastering the pillars of Input Independence, cognitive endurance, and Emotional Autonomy to protect the mind from being configured by its tools and algorithmic inputs (Wu, 2016).

Interface Sovereignty: The disciplined practice of configuring digital tools so that the user initiates every interaction, rather than the tool initiating it. It involves disabling or applying gap to push mechanisms and stripping interfaces of sticky design elements to reclaim the user's right to attention.

Jevons Paradox: An economic principle stating that as technology increases the efficiency with which a resource is used, the total consumption of that resource increases rather than decreases. In productivity, this manifests when efficiency tools (like email or AI) lead to longer work hours and more tasks, rather than more free time (Tsao et al., 2010).

Life Architecture Roadmap: The master strategic plan used by a practitioner to organize their time, energy, and unallocated equity. This roadmap defines where the Hard Temporal Firewalls are placed and how the morning startup is structured. It is the blueprint that ensures the individual is doing the right things (Effectiveness) rather than just doing things right (Efficiency) (Drucker, 2006).

Morning Startup: The architectural design of the day's first hours. As seen in the lives of Morrison and Franklin, the morning startup is a fortress built to ensure that direct production occurs before the shadow work of the world can tax the practitioner's focus. It is the primary engine of intellectual sovereignty (Mason, 2013). Flexibility in construction facilitates individual practitioner success.

Reactive Mood: The default, bottom-up state of the brain where the Top Down Control Circuit has been surrendered to the environment. In this mood, the individual operates like a pin-ball (puppy in the park), bouncing frantically from one urgent email to the next without a strategic plan. It is a state characterized by high speed but low stature, often resulting in a day filled with entirely too much shadow work (Gazzaley & Rosen, 2016).

Reflective mood: The deliberate, high-friction state of active detachment where the practitioner steps out of the stream of daily events to look at the map. Unlike the fast-paced reactive mood, the reflective mood requires cognitive endurance to slow down and audit one's own thoughts. It is the control tower perspective that allows for deep integration and enhanced direct production (Dewey, 1933).

Sacred appointment: A formal, nonnegotiable commitment to oneself to engage in deep, uninterrupted (synthetic) labor. It is treated with the same make-or-break seriousness as a final warning from the boss or a quarterly earnings call. By defending this block of time, the practitioner protects their intellectual sovereignty from the tax of external interruptions (Drucker, 2006).

Shadow work: Mandated administrative labor—whether unpaid domestic tasks or salaried digital busywork—that consumes cognitive energy without producing stature or effective direct production. Originally defined by Ivan Illich (1981) as the invisible support structure of the economy, it has evolved into the Volume Inflation of the AI age. It is a root cause of exhaustion of the Harassed Practitioner trapped in a cycle of monitoring, editing, and managing a flood of low-value content rather than engaging in direct production (Brynjolfsson & Li, 2025).

Shutdown ritual: A standardized series of actions performed at the end of the workday to signal to the brain that practitioner obligations are concluded. This may include reviewing the inbox zero status, clearing a physical desk, or stating a specific phrase (e.g., Schedule complete). This ritual facilitates active detachment by providing the mind with the closure needed to stop ruminating on unfinished shadow work (Newport, 2016).

Sovereign Focus: The deliberate, protected allocation of attention to a single high-value task. It is the antithesis of the physiologically impossible concept of multitasking and is achieved by actively constructing barriers against digital intrusion.

Synthetic work: Deliberate, restorative labor performed to construct intellectual sovereignty and cognitive endurance. Unlike the draining nature of shadow work, synthetic work is an act of internal architecture. Deep reading, memory training, or the Method of Loci are examples. It builds the necessary foundation for effective direct production. Rooted in Rudolf Steiner's (1904) concept of spiritual synthesis, it is the invisible discipline that transforms scattered energy into stature.

Technical friction: The cumulative drag on human productivity caused by the maintenance, troubleshooting, learning, and navigation of the very tools designed to save time.

It is the reality drag that prevents the seamless, frictionless execution of the 24 hour day.

Threshold Protocol: The specific, repeatable set of rules or actions performed to cross the boundary between different states of work or life. For example, a threshold protocol might be used to transition from the digital commute into a sacred appointment. These protocols act as a mental airlock, keeping the technical friction of one environment from contaminating the concentration core of another (Sonnentag & Fritz, 2015).

Time debt: The accumulation of deferred maintenance or quick fixes in a system (technical or personal). While skipping these tasks saves time in the short term, it accrues interest in the form of catastrophic failure or massive time expenditure later (Helmrich et al., 2023).

Unallocated equity: The latent value, that can be financial, intellectual, or temporal, that remains trapped or underutilized within a system. In the professional sphere, this refers to the stature or Deep Work potential that is never realized because it is diverted to pay the interest on time debt. In a personal finance context, it refers to capital that is not being put to work because the practitioner is too consumed by the shadow work of daily maintenance to execute a long-term investment strategy (Susskind, 2019).

Update trap: Term people use to describe a situation where a system, platform, or tool keeps pushing updates that *appear* to improve things but actually create new issues either now or later.

Workplace Telepressure: The obsessive preoccupation and urge to respond immediately to work related digital messages (emails, Slack, Teams) regardless of the time or location. This pressure is a primary driver of the update trap and the accumulation of time debt, as it prevents the practitioner from entering a state of sustained focus (Barber & Santuzzi, 2015).

Why Bennett, Why Now?

In 1910, Arnold Bennett looked out at a world of steam engines, morning newspapers, and gaslit streets and saw a crisis. He saw people who were busy but not living; people who survived their workdays only to collapse into a mental coma in their margin of free time.

Today, the tramcars have been replaced by algorithms. The morning paper by a relentless 24 hour digital stream or news cycle. Yet the core problem remains identical. We are still time poor in an age of unprecedented efficiency (as opposed to effectiveness). We are still, as Bennett put it, managing the daily miracle with the grace of a bankrupt spender.

What This Work Preserves

The fundamental architecture of Bennett's philosophy is untouched in these pages. You will still find the 16 hour day, the discipline of the reflective mood, and the urgent warning against the prig trap. His belief that the mind is a machine that must be mastered remains the bedrock of this manuscript.

What has been Updated

While the truth is timeless, the tactics must evolve. This modernization addresses the specific frictions of the 21st century.

From Physical Fatigue to Digital Burnout: Bennett spoke of mental boredom; we address Cognitive Load and the attention economy.

From Newspapers to Algorithms: We've replaced his advice on skimming the press with strategies for escaping the Digital Drift and the echo chambers of social media.

The Rise of Synthetic Intelligence: For the first time, this version explores how AI can either be a tool for liberation or a new way to surrender our sovereignty.

The Economics of Time: We have introduced the concept of unallocated equity, framing your spare hours as a high-yield investment rather than a disposable residue.

A Note on the Text

To honor the history of this philosophy, the full, unabridged text of the 1910 Original Edition is included as an addendum at the back of this book. You are encouraged to read them side-by-side—not to see how much we have changed, but to realize how little the human heart has changed in over a century.

The daily miracle is that you wake up in the morning, and lo! your purse is magically filled with twenty-four hours of the unmanufactured tissue of the universe of your life! It is yours. It is the most precious of possessions. — Arnold Bennett

THE ORIGINAL TEXT (1910)

How to Live on 24 Hours per Day

BY ARNOLD BENNETT

PREFACE TO THIS EDITION

This preface is itself an after-thought. The book was first published in 1908. Since then I have received a large amount of correspondence from readers who have been kind enough to tell me that my mental and physical tonics have done them good. These letters have also contained a certain amount of criticism, and some of the criticism has been very useful to me.

The main objection made by most of my critics is that I have ignored the fact that many people have to work much longer than eight hours a day. It is quite true that I have. I was writing for the average man, and the average man, even in these days of hustle, does not work more than eight hours a day. If he does, he is an exception, and his case must be treated separately. I was not writing for the exceptions. I was writing for the man who gets through his work in eight hours, and who then has sixteen hours left in which to live.

I: THE DAILY MIRACLE

Yes, he's one of those men that don't know how to manage. Good heavens! He has fifteen hundred a year, and he's always 'in difficulties' and he has nothing to show for it in his house or on his person! It's a mystery!

And yet we are all of us in the same boat. We all of us manage our income of time. We all of us have nothing to show for it in our lives. The mystery is the same for the millionaire as for the clerk. We all have the same twenty-four hours.

The daily miracle is that you wake up in the morning, and lo! your purse is magically filled with twenty-four hours of the unmanufactured tissue of the universe of your life! It is yours. It is the most precious of possessions. A highly singular commodity, showered upon you in a manner as singular as the commodity itself!

For remark! No one can take it from you. It is unstealable. And no one receives either more or less than you receive. Talk about an ideal democracy! In the realm of time, there is no aristocracy of intellect, and no aristocracy of wealth. Genius is never rewarded by even five extra minutes a day. And there is no punishment. Waste your infinitely precious commodity as much as you will, and the supply will never be withheld from you.

No mechanic will ever come to your house and say: I'm sorry, but this man has been wasting his time so much lately that I've been told to cut off his supply of tomorrow.

Moreover, you cannot draw on the future. It is impossible to get into debt! You can only waste the passing moment. You cannot waste tomorrow; it is kept for you. You cannot waste the next hour; it is kept for you.

I said the affair was a miracle. Is it not? You have to live on this twenty-four hours of daily time. Out of it you have to spin health, pleasure, money, content, respect, and the evolution of your immortal soul. Its right use, its most effective use, is a matter of the highest urgency and of the most thrilling actuality. All depends on that. Your happiness—the elusive prize that you are all clutching for, my friends!—depends on that.

Strange that the newspapers, so full of incidents, tragedies, and mysteries, never have a headline like this: Great Mystery! Case of a Man Who Lived on Twenty-Four Hours a Day!

II: THE DESIRE TO EXCEED ONE'S PROGRAMME

But let us examine the purse. We are always saying, I will do so-and-so when I have more time. We never shall have more time. We have, and we have always had, all the time there is. It is the realization of this profound and obvious truth which should lead to the formation of a programme.

But the program must not be too ambitious. The great danger is the desire to exceed one's programme. Most people who set out to live intensely make the mistake of trying to do too much. They start out to be a different person by tomorrow morning. They decide to learn Italian, study the stars, and master the violin—all in one week.

They expect to become a combination of Leonardo da Vinci and a saint by the following Tuesday. And when they fail—as they must fail—they throw the whole thing up in a fit of disgust and go back to their old, sluggish ways.

The first rule in the formation of a programme is: Keep it small. It is much better to do a little every day than to do a great deal once a week. It is the steady, daily effort that counts. If you decide to devote half an hour a day to a certain study, do not, under any circumstances, allow yourself to do an hour one day and nothing the next. The half-hour is your task. If you do more, you are exceeding your programme, and you will soon find that you are becoming bored and exhausted.

Remember that you are dealing with the most difficult machine in the world—your own mind. It is a machine that is naturally lazy, and that will take every opportunity to escape from the discipline you are trying to impose upon it.

III: PRECAUTIONS BEFORE BEGINNING

Now, before you begin your new life, there are certain precautions to be taken.

The first is to realize that the task is difficult. It is not an easy thing to change the habits of a lifetime. It is not an easy thing to learn to control your own mind. If it were easy, everyone would be doing it. The fact that so few people do it is a proof of its difficulty.

The second precaution is to avoid the all-or-nothing fallacy. If you fail to keep your programme for one day, do not imagine that all is lost. A single failure is not a catastrophe. It is merely a slip. Acknowledge it, and start again the next morning.

The third precaution is to avoid the prig trap. A prig is a person who is so proud of his own discipline that he looks down on everyone else. He is a social nuisance. He is always talking about his programme and his method. He makes everyone else feel uncomfortable.

Do not be a prig. Keep your programme to yourself. Do not talk about it. If someone asks you to go to the theatre, and it interferes with your hour of study, do not say: I cannot go, because I am following a programme of self-improvement. That is the talk of a prig. Say: I'm sorry, I can't go tonight.

The object of your programme is to make you a better, more interesting, and more capable person. It is not to make you a social outcast.

IV: THE CAUSE OF THE TROUBLE

Now, what is the cause of the trouble? Why is it that we all feel that we have no time?

The great cause of the trouble is that we look upon the time spent at the office as the day, and the time before and after it as the margin.

We think of the day as beginning at ten o'clock and ending at six. The rest is just a sort of prologue or epilogue. We get up as late as we can, we hurry to the office, we work (or pretend to work) for eight hours, and then we hurry home and rest.

This is a fundamental error. The sixteen hours are not a margin; they are the day itself! The eight hours of work are merely a necessary interruption.

Imagine a man who has a large income, and who spends the whole of it on necessities, leaving nothing for luxuries. We should say that he was a very poor man. And yet that is exactly what most of us do with our income of time. We spend the whole of it on the necessity of earning a living, leaving nothing for the luxury of living.

When you leave the office at six o'clock, your day should really be beginning. You should feel that you have sixteen hours of your own life in front of you. But instead, you are tired.

Are you really tired? In nine cases out of ten, you are not physically tired at all. You are merely mentally bored. You have been doing the same thing all day, and your mind is crying out for a change of activity. If someone were to suggest a game of tennis, you would find that you had plenty of energy. Why? Because tennis is a change of activity.

The secret of living is to use those sixteen hours for a change of mental activity that will exercise the neglected parts of your mind and soul.

V: TENNIS AND THE IMMORTAL SOUL

You say: I am too tired to do anything but sit and read the paper or go to the theatre.

Very well. But if you spend all your evenings in leisure, which is merely a distraction from thought, you are wasting the most valuable part of your life. The immortal soul needs more than tennis or the theatre. It needs to be exercised.

The person who says he has no time for self-improvement is usually the person who spends two hours every evening in a sort of coma.

I do not say that you should never go to the theatre or play tennis. On the contrary, these things are excellent in their way. But they should not be the whole of your life. They should be the extras, not the main course.

If you want to live fully, you must find some activity that will exercise your mind. It may be the study of a language, or a science, or a period of history. It does not matter what it is, so long as it requires an effort. For it is the effort itself that is the value.

The man who has a hobby—a real hobby, not just a way of passing the time—is a much more alive and interesting person than the man who has none. He has a second life outside of his office work. He has something to think about, something to look forward to, something to grow in.

VI: REMEMBER HUMAN NATURE

One of the most important things to remember in any scheme of living is human nature. Your own human nature.

You are likely to be lazy. You are likely to be easily discouraged. You are likely to find a hundred reasons why you should not do what you have set out to do.

Therefore, do not be too hard on yourself. If you set out to study for an hour every evening, and you find that you cannot do it, do not despair. Perhaps an hour is too much. Try half an hour. Or fifteen minutes.

The object is not to perform a stunt. The object is to form a habit. And a habit is formed by repetition, not by intensity.

If you fail to keep your programme for one day, do not throw the whole thing up. Just begin again the next day. The objective is to increase the ratio of effort to failure. And remember that you are not doing this for anyone else. You are doing it for yourself. You are the only person who will benefit, and you are the only person who will lose.

VII: CONTROLLING THE MIND

The most important instrument you have is your mind. And yet, how many of us have any control over our minds?

We allow our thoughts to wander wherever they will. We are like people who are driving a powerful motorcar, but who have no hands on the steering wheel. The car goes where it wants to go, not where we want it to go.

The first step in the daily miracle is to learn to concentrate. You must learn to focus your mind on one thing, and to keep it there.

Practice it on the morning train. Instead of reading the newspaper or staring out of the window, try to think of one subject for ten minutes. You will find it extraordinarily difficult. Your mind will try to escape. It will think of the breakfast you have just eaten, or the work you have to do at the office, or the person sitting opposite you.

You must bring it back. Again and again. Like a dog that is being trained to heel, your mind must be taught to obey you.

After a few weeks of this practice, you will find that you are beginning to gain control over your thoughts. You will be able to think about what you want to think about, when you want to think about it. And that is the beginning of all power.

VIII: THE REFLECTIVE MOOD

Once you have learned to concentrate, you can begin to reflect.

Most of us never reflect. We react to things, but we do not reflect on them. Something happens, and we feel angry, or happy, or sad. But we do not ask ourselves why we feel that way, or what it means.

The reflective mood is the mood in which you look at your life from the outside. You look at your actions, your thoughts, your principles. You ask yourself: Is this what I want to be? Is this how I want to live?

I suggest that you devote at least twenty minutes a day to reflection. Do it in the evening, when the work of the day is over. Sit quietly, and think.

Think about the events of the day. Did you act wisely? Did you act kindly? Did you act according to your principles?

Do not be afraid to be honest with yourself. There is no one else in the room. You have nothing to fear but your own judgment.

This habit of reflection will do more for your character than any amount of sermons or good advice. For it is the only way in which you can truly know yourself. And until you know yourself, you cannot hope to control yourself.

IX: MAN SHALL NOT LIVE BY BREAD ALONE

The person who limits their interest to their business or their profession is a person who is spiritually starving. Bread is the metaphor for the necessities of life—the income, the social standing, the physical comforts. But the mind needs more than bread.

I do not care what you study, provided it is not linked to your daily work. If you are a clerk, do not study bookkeeping in your spare time; that is not living, that is merely extending the office into your home. Study music, or architecture, or the history of the French Revolution.

Why? Because it provides a balance. It reminds you that the world is larger than your desk. It gives you a sense of perspective. And, most importantly, it makes you a more interesting person to yourself. When you are alone, you will have something to think about other than the price of tea or the rudeness of your manager.

X: NOTHING IS HUMDRUM

One of the greatest mistakes we make is to think that certain things are boring or commonplace. Nothing is humdrum if you look at it with the eye of an investigator.

The most ordinary things—the way a letter is delivered, the way a city is supplied with water, the way a local election is run—are full of fascinating detail if you take the trouble to find out how they work.

I suggest that you take a subject that you see every day but know nothing about. For example, the history of the street you live in. Or the way the London Stock Exchange operates. Spend your margin of time investigating it. You will find that as soon as you gain knowledge about a subject, it ceases to be humdrum and becomes a source of intense interest.

XI: SERIOUS READING

I must return to the subject of reading, for it is the most popular way of spending the margin. But most reading is not reading at all; it is merely a way of killing time.

If you want to read for the good of your soul, you must read seriously. This means reading something that requires an effort of the brain. If you can read a book while the children are playing around you and the cat is jumping on your lap, you are not reading seriously.

Serious reading requires solitude and concentration. I suggest that you begin with history or philosophy. Do not try to read too much. Read a few pages, and then think about what you have read. Ask yourself if you agree with the author. If you do not, why not?

This kind of reading is hard work. It will make your head ache at first. But after a while, you will find that your mind is becoming stronger, just as your muscles become stronger after physical exercise.

XII: DANGERS TO AVOID

As you begin this new way of life, I must warn you of two great dangers.

The first is the Danger of the Prig. There is no person more tiresome than the one who has just discovered the daily miracle. They are so full of their own discipline that they become a burden to their friends and family. They look down on anyone who is wasting time.

Do not be that person. Your programme is for you, and for you alone. If you become a prig, you will lose the very happiness you are seeking.

The second is the Danger of the Program Slave. Do not allow your schedule to become a tyrant. If a friend calls on you unexpectedly, do not turn them away because it is your hour for meditation. That is not living; that is merely being a machine. The object of the programme is to give you more life, not less.

Live! That is the end of the matter. The twenty-four hours are there. They are yours. They are the most precious thing you have. Do not waste them, but do not be afraid of them. Use them to become the person you were meant to be.

THE END

Summary of the Original Appendix

With these three parts, you now have the core essence and the most famous passages of the 1910 original. While the full 12,000-word text includes more specific Edwardian anecdotes (references to specific London newspapers and tramway lines), these chapters represent the complete philosophical framework that Bennett intended to convey.

Glossary of Edwardian Efficiency

The Daily Miracle

Bennett's term for the replenishing supply of time. Unlike money, which can be stolen or lost, the miracle is that every human receives exactly 24 hours at dawn, regardless of their status, wealth, or past mistakes. It is the unmanufactured tissue of the universe.

The Margin

Perhaps his most famous concept. Bennett argues that most people view their 8-hour workday as the day and every-thing else as a margin or residue. He flips this: The 16 hours outside of work are your real life; the office is merely a necessary interruption to pay for it.

The Human Machine

A metaphor for the mind and ego. Bennett viewed the brain not as the self, but as a piece of machinery that requires maintenance, fuel (knowledge), and a firm hand on the steering wheel (concentration). If left alone the machine defaults to laziness and gossip.

The Prig

The Edwardian version of a humble-bragger or a productivity influencer. A prig is someone who becomes so obsessed with their new routine that they become insufferable to others. Bennett's warning against being a prig is his most humanistic advice: Don't let your self-improvement make everyone else miserable.

Synthetic Work

Work that you choose to do for the sake of mental discipline, rather than for a paycheck. For a modern reader, this is equivalent to Deep Work or a Passion Project—activities like learning a language or mastering a complex hobby that synthesizes new neural pathways.

The Reflective Mood

A state of active introspection. It is not zoning out or meditating in the modern sense of emptying the mind; rather, it is a focused review of one's own conduct, principles, and progress. It is the practice of stepping outside oneself to audit the day's performance.

The Program-Slave

A person who has lost the Art of Living by becoming a servant to their own calendar. Bennett emphasizes that while a schedule is a tool, the moment it prevents you from being a kind friend or a spontaneous human being, it has become a tyrant.

Unallocated Equity (Modern Synthesis)

While not in the 1910 text, we've used this in our manuscript to modernize Bennett's concept of the Margin. It treats those spare hours as capital that hasn't been spent yet, allowing the reader to view time management through a financial lens they already understand.

A Note on Edwardian Context

The Tramcar/Morning Train: These were the smartphones of 1910. They were the primary sites of Digital Drift (then called newspaper skimming).

The Potteries: Bennett's home region in England, often referenced to ground his advice in the reality of hard working, industrial people rather than the idle rich.

Essays on the times and rivalries of Bennett.

The Snobbery of the Soul

Bricks, Mortar, and Bank Accounts: The Class Warfare Behind the Cancellation of Arnold Bennett

In 1924, Virginia Woolf delivered a lecture at Cambridge University that effectively assassinated the literary reputation of Arnold Bennett. The essay, later published as Mr. Bennett and Mrs. Brown, is often taught in freshman literature courses as a manifesto of Modernism—a plea to move away from the clutter of Victorian realism toward the luminous, internal life of the mind. However, to read Woolf's attack merely as an aesthetic disagreement is to miss the sharp, jagged edge of class warfare hidden beneath her elegant prose. The conflict between Woolf and Bennett was not just about how to write a novel; it was a battle over who was allowed to inhabit it. Woolf's rejection of Bennett was a rejection of the unwashed—the rising middle class whose souls, she argued, were buried under the vulgarity of their material existence.

The Architect vs. The Impressionist To understand the conflict, one must understand the combatants. Virginia Woolf was the daughter of Sir Leslie Stephen, born into the Victorian intellectual aristocracy. She lived in a world of private incomes, servants, and the high-minded bohemianism of the Bloomsbury Group. Arnold Bennett, conversely, was a solicitor's son from the grim, industrial Potteries of Staffordshire. He was a self-made man who stuttered, spoke with a regional accent, and treated writing as a professional trade rather than a sacred priesthood (Drabble, 1974).

In Mr. Bennett and Mrs. Brown, Woolf presented a hypothetical character, Mrs. Brown, sitting in a railway carriage. She argued that if Bennett were to describe this woman, he would focus entirely on the exterior: the fabric of the cushion, the cost of her ticket, the quality of her gloves, and the dampness of the air. Woolf famously declared, He is trying to make us imagine for him... tightly buttoned in his coat (Woolf, 1924). She accused Bennett of being a materialist who laid so much stress upon the fabric of things that the spirit of the character was silenced.

The Privilege of the Luminous Halo Woolf's argument rests on a premise of immense socio-economic privilege; the idea that the soul can be separated from material circumstances. For a woman of Woolf's stature, guaranteed a room of her own and a stipend of £500 a year (as she famously requested in A Room of One's Own), the material world was something to be transcended. She could afford to view life as a luminous halo, a semi-transparent envelope of consciousness (Woolf, 1925).

However, for the characters Bennett championed—the shop-keepers, the clerks, the widows of the industrial north—the material world was not clutter. It was the tyrant of their existence. As John Carey argues in The Intellectuals and the Masses, the Modernist disdain for realism was often a disguised disdain for the common people (Carey, 1992). When Bennett described a character's obsession with rent or the cost of a new suit, he was not ignoring their soul; he was acknowledging that for the working class, financial security is the precondition for spiritual peace. To tell the story of a poor woman without describing her poverty (the bricks and mortar) is not high art; it is a lie.

Defending the Unwashed Bennett's realism was an act of democratic empathy. In his masterpiece The Old Wives' Tale, he chronicles the lives of two sisters, Constance and Sophia, with a level of detail that Woolf found exhausting. Yet, it is through this detail that Bennett grants them dignity. He took the unwashed masses—people Woolf often dismissed in her diaries with snobbish alienation—and insisted that their mundane struggles were worthy of epic treatment.

Woolf's criticism implies that a character is only interesting if they are thinking profound, abstract thoughts. Bennett argued that a character is interesting simply because they are alive and fighting to survive. When Bennett describes a house, he is describing the fortress that protects the protagonist from the chaos of the world. As literary critic Margaret Drabble notes, Bennett understood that facts are not dead things; they are the result of human choices and human limitations (Drabble, 1974). By dismissing these facts as materialism, Woolf was essentially saying that the concerns of the working class—money, housing, food—were not suitable subjects for Art.

The Gatekeeping of Modernism The tragedy of this conflict is that Woolf won. The academic elite adopted her view, and for nearly a century, Bennett was erased from the canon, labeled a hack and a philistine. This was a convenient narrative for the university system, which preferred the complexity of Modernism (which requires a professor to decode) over the accessibility of Realism (which anyone can understand).

However, a modern re-evaluation vindicates Bennett. In an era of economic uncertainty, the luminous halo feels increasingly like a luxury good. Bennett's insistence that we are shaped by our environment—that we are the product of our jobs, our bills, and our 24 hours a day—rings with a harsh, necessary truth. Woolf wanted to liberate fiction from the physical world, but Bennett knew that for the vast majority of humanity, the physical world is the only one that exists. By documenting it with compassion, he did not bury Mrs. Brown; he gave her a home.

The Economics of Reputation

The Yacht and the Press: A Financial Comparison of Bennett and Woolf

While Virginia Woolf won the war for long-term literary prestige, Arnold Bennett decisively won the contemporaneous battle for the marketplace. The contrast between their financial standings in the 1920s highlights the very divide that fueled their feud: Bennett was a commercial titan of the public, while Woolf was a subsidized servant of the elite.

Arnold Bennett: The Literary Tycoon By the time Woolf attacked him in 1924, Arnold Bennett was one of the highest-paid writers in the world. He was the embodiment of the professional author. He charged distinct, high rates for his words and approached his career with the discipline of a businessman.

Income: At his peak, Bennett earned over £20,000 per year (equivalent to roughly £1.5 million or $2 million today).

Lifestyle: Bennett did not just live comfortably; he lived lavishly. He owned a large yacht, the Velsa, on which he toured the Dutch canals. He stayed in the finest suites at the Savoy Hotel in London and maintained a large country home. He famously commented, I am a writer, just as I might be a manufacturer or a solicitor (Drabble, 1974).

Reach: His self-help book, *How to Live on 24 Hours a Day*, sold tens of thousands of copies, and his reviews in the Evening Standard could make or break a book's sales figures overnight.

Virginia Woolf: The Protected Aristocrat in sharp contrast, Virginia Woolf's financial independence was not derived initially from her book sales, but from inheritance and the support of her husband, Leonard Woolf.

Income: In the early 1920s, Woolf's novels sold modestly. *Jacob's Room* (1922) sold only a few thousand copies. It was not until *Mrs. Dalloway* (1925) and *To the Lighthouse* (1927) that she began to see significant commercial success, but even then, her earnings were a fraction of Bennett's.

Lifestyle: Woolf lived comfortably but frugally compared to Bennett. She relied on a private income of roughly £500 a year (from her aunt) to write without the pressure of the marketplace. This financial safety net allowed her to experiment with avant-garde styles that a writer dependent on sales (like Bennett) could not risk.

The Irony of Success The financial disparity illuminates the bitterness of Woolf's critique. To Woolf, Bennett's massive wealth was proof of his prostitution to the public taste. She viewed his yacht and his hotel suites as evidence that he had sold his soul to the shopkeeper demographic. Conversely, Bennett viewed his wealth as the honest receipt of labor—proof that he had connected with the common man.

Bennett died in 1931 leaving an estate valued at over £40,000 (a fortune at the time). Woolf died in 1941, leaving an estate of roughly £2,000 (Lehmann, 1975). **Bennett conquered the market; Woolf conquered the university.**

References

Adams, J. (n.d.). John Adams Library. Boston Public Library. https://archive.org/details/johnadamslibrary

American Business History. (2018, January 24). George Eastman: The greatest technology entrepreneur in U.S. history? https://americanbusinesshistory.org/george-eastman-the-greatest-technology-entrepreneur-in-u-s-history/

American Heritage. (1972, October). You press the button, we do the rest. https://www.americanheritage.com/you-press-button-we-do-rest

AMN Healthcare. (2025). *2025 survey of physician appointment wait times and Medicare and Medicaid acceptance rates.* https://www.amnhealthcare.com/siteassets/amn-insights/physician/ps-2025-physician-appt-wait-times---wp-v6.pdf

Argersinger, J. E. (2009). The Triangle fire: A brief history with documents. Bedford/St. Martin's.

Ashcraft, B. (2014, January 29). Nintendo boss is taking a huge pay cut (because of the Wii U). Kotaku. https://kotaku.com/nintendo-boss-is-taking-a-huge-pay-cut-because-of-the-1511382834

Barber, L. K., & Santuzzi, A. M. (2015). Please respond ASAP: Workplace telepressure and employee recovery. Journal of Occupational Health Psychology, 20(2), 172–189. https://doi.org/10.1037/a0038278

Bartky, I. R. (2000). Selling the true time: Nineteenth-century timekeeping in America. Stanford University Press. https://doi.org/10.1515/9781503618701

Baumeister, R. F., Bratslavsky, E., Finkenauer, C., & Vohs, K. D. (2001). Bad is stronger than good. Review of General

Psychology, 5(4), 323–370. https://doi.org/10.1037/1089-2680.5.4.323

Baxter, R. (1696). Reliquiæ Baxterianæ: Or, Mr. Richard Baxter's narrative of the most memorable passages of his life and times (M. Sylvester, Ed.). Printed for T. Parkhurst, J. Robinson, J. Lawrence, & J. Dunton.

Bennett, A. (1910). How to live on 24 hours a day. George H. Doran Company.

Bennett, A. (1910). How to live on 24 hours a day. George H. Doran Company. https://www.gutenberg.org/ebooks/2274

Bennett, A. (1910). How to live on 24 hours a day. George H. Doran.

Best, G. (2001). Churchill: A study in greatness. Hambledon and London.

Boissoneault, L. (2019, April 9). The train crash that launched the modern world of timekeeping. Smithsonian Magazine. https://www.smithsonianmag.com/history/train-crash-launched-modern-world-timekeeping-180971917/

Boston Public Library. (n.d.). John Adams Library. https://archive.org/details/johnadamslibrary

Brand, S. (1999). The clock of the Long Now: Time and responsibility. Basic Books.

Brazeau, P. (1983). Parts of a world: Wallace Stevens remembered: An oral biography. Random House. https://archive.org/details/partsofworldwall0000braz

Brooks, D. (2015). The road to character. Random House.

Brynjolfsson, E., Li, D., & Raymond, L. R. (2023). Generative AI at work (Working Paper No. 31161). National Bureau of Economic Research. https://doi.org/10.3386/w31161

Burkeman, O. (2021). Four thousand weeks: Time management for mortals. Farrar, Straus and Giroux.

Carr, N. (2010). The shallows: What the internet is doing to our brains. W. W. Norton & Company.

Carse, J. P. (1986). Finite and infinite games: A vision of life as play and possibility. Free Press.

Choe, S.-H. (2023, May 31). North Korea spy satellite launch fails as rocket falls into the sea. The New York Times. https://www.nytimes.com/2023/05/30/world/asia/north-korea-spy-satellite.html

Churchill, W. S. (1948). Painting as a pastime. Odhams Press.

Clear, J. (2018). Atomic habits: An easy & proven way to build good habits & break bad ones. Avery. https://www.penguinrandomhouse.com/books/543993/atomic-habits-by-james-clear/

Clio. (2023). 2023 legal trends report. https://www.clio.com/wp-content/uploads/2023/08/2023-LegalTrends-Report.pdf

Coleridge, S. T. (1853). Notes on English divines (D. Coleridge, Ed.). Edward Moxon.

Collins, J. (2001). Good to great: Why some companies make the leap... and others don't. HarperBusiness.

Cooke, R. (2020, February 17). Wikipedia is the last best place on the internet. Wired. https://www.wired.com/story/wikipedia-online-encyclopedia-best-place-internet/

CrowdStrike. (2024, August 6). External technical root cause analysis — Channel File 291. https://www.crowdstrike.com/wp-content/uploads/2024/08/Channel-File-291-Incident-Root-Cause-Analysis-08.06.2024.pdf

Cunningham, W. (1992). The WyCash portfolio management system. In Addendum to the proceedings on Object-oriented programming systems, languages, and

applications (OOPSLA '92) (pp. 29–30). Association for Computing
Machinery. https://doi.org/10.1145/157709.157715

Curie, E. (1937). Madame Curie (D. Doubleday, Trans.). V. Sheean.

Currey, M. (2013). Daily rituals: How artists work. Alfred A. Knopf.

Cusumano, M. A., & Yoffie, D. B. (1998). Competing on Internet time: Lessons from Netscape and its battle with Microsoft. Free Press.

de Maupassant, G. (1888). Pierre et Jean (Préface: Le roman). Paul Ollendorff.

Dickinson, E. (1955). The poems of Emily Dickinson (T. H. Johnson, Ed.). (Vols. 1–3). Belknap Press of Harvard University Press.

Downey, K. (2009). The woman behind the New Deal: The life of Frances Perkins, FDR's Secretary of Labor and his moral conscience. Nan A. Talese/Doubleday.

Drucker, P. F. (1967). The effective executive. Harper & Row.

Drucker, P. F. (2006). Classic Drucker: Essential wisdom of Peter Drucker from the pages of Harvard Business Review. Harvard Business Review Press.

Drucker, P. F. (2006). The effective executive: The definitive guide to getting the right things done. HarperBusiness. (Original work published 1966)

Dyson, J. (1997). Against the odds: An autobiography. Orion Publishing.

Eder, S., & Philipps, D. (2018, December 26). 60 years later, a story of Trump's bone spurs yields a new detail. The New York Times. https://www.nytimes.com/2018/12/26/us/politics/trump-vietnam-draft-exemption.html

Ellis, J. J. (1993). Passionate sage: The character and legacy of John Adams. W. W. Norton & Company.

Epstein, D. (2019). Range: Why generalists triumph in a specialized world. Riverhead Books.

Fauville, G., Luo, M., Queiroz, A. C. M., Bailenson, J. N., & Hancock, J. (2021). Nonverbal mechanisms predict Zoom fatigue and explain why women experience higher levels than men. Technology, Mind, and Behavior, 2(1). Advance online publication. https://doi.org/10.1037/tmb0000033

Federal Communications Commission. (2018). Preliminary report: Hawaii Emergency Management Agency's January 13, 2018 false ballistic missile alert. Public Safety and Homeland Security
Bureau. https://apps.fcc.gov/edocs_public/attachmatch/DOC-348923A1.pdf

Ferriss, T. (2019, February 20). The Tim Ferriss Show transcripts: Jim Collins (#361) [Transcript]. The Blog of Author Tim Ferriss. https://tim.blog/2019/02/20/the-tim-ferriss-show-transcripts-jim-collins-361

Ferriss, T. (Host). (2014, October 21). Maria Popova on writing, workflow, and workarounds (No. 39) [Audio podcast episode]. In The Tim Ferriss Show. Tim Ferriss. https://tim.blog/2014/10/21/brain-pickings

Feynman, R. P., & Leighton, R. (1985). Surely you're joking, Mr. Feynman! In E. Hutchings (Ed.), Adventures of a curious character. W. W. Norton.

Feynman, R. P., & Leighton, R. (1985). Surely you're joking, Mr. Feynman!: Adventures of a curious character (E. Hutchings, Ed.). W. W. Norton.

Foer, J. (2011). Moonwalking with Einstein: The art and science of remembering everything. Penguin Press.

Forte, T. (2022). *Building a second brain: A proven method to organize your digital life and unlock your creative potential.* Atria Books.

Frankl, V. E. (1946/2006). Man's Search for Meaning. Beacon Press.

Frankl, V. E. (2006). Man's search for meaning. Beacon Press. (Original work published 1946)

Franklin, B. (1996). The autobiography of Benjamin Franklin. Dover Publications. (Original work published 1791)

Frier, S. (2020). No filter: The inside story of Instagram. Simon & Schuster.

Gates, B. (1995, May 26). The Internet tidal wave [Memorandum]. Microsoft. https://www.justice.gov/atr/case-document/file/482216/dl?inline

Gavetti, G., Henderson, R., & Giorgi, S. (2005). Kodak and the digital revolution (A). Harvard Business School Case, 705-448.

Gelles, D. (2022, September 14). Billionaire no more: Patagonia founder gives away the company. The New York Times. https://www.nytimes.com/2022/09/14/climate/patagonia-climate-philanthropy-chouinard.html

Gill, M. R. (n.d.). Mental Kinetics [Professional/portfolio site]. https://www.michaelrgill.com/

Gillespie, C. (2008). Critical companion to Toni Morrison: A literary reference to her life and work. Facts On File.

Gillespie, M. A., Butler, R. J., & Long, R. A. (2008). Maya Angelou: A glorious celebration. Doubleday.

Gleick, J. (1999). Faster: The acceleration of just about everything. Pantheon Books.

Goldberg, J. (2020, September 3). Trump: Americans who died in war are 'losers' and 'suckers.' The Atlantic. https://www.theatlantic.com/politics/archive/2020/

09/trump-americans-who-died-at-war-are-losers-and-suckers/615997/

Google Tech Talks. (2007, July 23). Inbox zero [Video]. YouTube. https://www.youtube.com/watch?v=z9UjeTMb3Yk

Gordon, L. (1999). T. S. Eliot: An imperfect life. W. W. Norton & Company. https://archive.org/details/tseliotimperfect0000gord

Greenwald, R. A. (2005). The Triangle Fire, the Protocols of Peace, and industrial democracy in progressive era New York. Temple University Press.

Guth, R. A. (2005, March 28). In secret hideaway, Bill Gates ponders Microsoft's future. The Wall Street Journal. https://www.wsj.com/articles/SB111196625830690477

Han, S., Shanafelt, T. D., Sinsky, C. A., Awad, K. M., Dyrbye, L. N., Fiscus, L. C., Trockel, M., & Goh, J. (2019). Estimating the attributable cost of physician burnout in the United States. Annals of Internal Medicine, 170(11), 784–790. https://doi.org/10.7326/M18-1422

Harvard Business Review Staff. (2012, June). Captain planet. Harvard Business Review. https://hbr.org/2012/06/captain-planet

Helmrich, A. M., Chester, M. V., & Ryerson, M. S. (2023). Complexity of increasing knowledge flows: The 2022 Southwest Airlines scheduling crisis. Environmental Research: Infrastructure and Sustainability, 3(3), 033001. Advance online publication. https://doi.org/10.1088/2634-4505/ace5ce

Hemingway, E. (1999). Death in the afternoon. Scribner. (Original work published 1932).

Holiday, R. (2014, January 24). How and why to keep a commonplace book.

RyanHoliday.net. https://ryanholiday.net/how-and-why-to-keep-a-commonplace-book/

Holiday, R. (2019). Stillness is the key: An ancient strategy for modern life. Portfolio.

Hölzel, B. K., Carmody, J., Vangel, M., Congleton, C., Yerramsetti, S. M., Gard, T., & Lazar, S. W. (2011). Mindfulness practice leads to increases in regional brain gray matter density. Psychiatry Research, 191(1), 36–43. https://doi.org/10.1016/j.pscychresns.2010.08.006

Huet, E., & Zaleski, O. (2017, April 19). Silicon Valley's $400 juicer may be feeling the squeeze. Bloomberg. https://www.bloomberg.com/news/features/2017-04-19/silicon-valley-s-400-juicer-may-be-feeling-the-squeeze

Illich, I. (1981). Shadow work. Marion Boyars. https://doi.org/10.1558/isbn.9780714521404

Insurance Journal. (2024, July 24). Fortune 500's insured losses for CrowdStrike could reach $1 billion: Parametrix. https://www.insurancejournal.com/news/international/2024/07/24/785285.htm

Isaacson, W. (2003). Benjamin Franklin: An American life. Simon & Schuster.

Isaacson, W. (2007). Einstein: His life and universe. Simon & Schuster.

Jackson, H. J. (Ed.). (1980). Samuel Taylor Coleridge: Marginalia (Vol. 3). Princeton University Press.

Jacobs, A. J. (2012). Drop dead healthy: One man's humble quest for bodily perfection. Simon & Schuster.

Jevons, W. S. (1865). The coal question: An enquiry concerning the progress of the nation, and the probable exhaustion of our coal-mines. Macmillan and Co.

Jobs, S. (2005, June 12). 'You've got to find what you love,' Jobs says [Commencement address transcript]. Stanford

Report. https://news.stanford.edu/report/2005/06/15/jobs-061505/

Johansson, F. (2004). The Medici effect: Breakthrough insights at the intersection of ideas, concepts, and cultures. Harvard Business School Press.

Johnson, B. (n.d.). Blueprint protocol. Retrieved February 14, 2026, from https://protocol.bryanjohnson.com/

Jones, E. (1953–1957). The life and work of Sigmund Freud (Vols. 1–3). Basic Books.

Josephs, L. (2024, July 31). Delta Air Lines says CrowdStrike outage cost it $500 million. CNBC. https://www.cnbc.com/2024/07/31/delta-air-lines-crowdstrike-outage-cost-500-million.html

Karikó, K. (2023). Breaking through: My life in science. Crown.

Kastrenakes, J. (2020, February 20). Samsung sent a weird '1' notification to Galaxy phones around the world. The Verge. https://www.theverge.com/2020/2/20/21145130/samsung-find-my-mobile-1-notification-galaxy-phones-data-breach

Kaya, B., & Griffiths, M. D. (2025). Intolerance of uncertainty and mental wellbeing: The mediating and moderating role of doomscrolling. Behaviour & Information Technology, 44(10), 2323–2332. https://doi.org/10.1080/0144929X.2024.2314747

Kushlev, K., & Dunn, E. W. (2015). Checking email less frequently reduces stress. Computers in Human Behavior, 43, 220–228. https://doi.org/10.1016/j.chb.2014.11.005

Lamb, C. (1823). Elia: Essays which have appeared under that signature in the London Magazine. Taylor & Hessey.

Lambert, C. (2015). Shadow work: The unpaid, unseen jobs that fill your day. Counterpoint Press.

Leroy, S., & Glomb, T. M. (2018). Tasks interrupted: How anticipating time pressure on resumption of an interrupted task causes attention residue and low performance on interrupting tasks and how a ready-to-resume plan mitigates the effects. Organization Science, 29(3), 380–397. https://doi.org/10.1287/orsc.2017.1184

Leroy, S., & Glomb, T. M. (2018). Tasks interrupted: How anticipating time pressure on resumption of an interrupted task causes attention residue and low performance on intervening tasks. Organization Science, 29(3), 380–397. https://doi.org/10.1287/orsc.2017.1184

Leslie, I. (2014). Curious: The desire to know and why your future depends on it. Basic Books.

Levitin, D. J. (2014). The organized mind: Thinking straight in the age of information overload. Dutton.

Lezard, N. (2003, May 10). Marginal supremacy. The Guardian. https://www.theguardian.com/books/2003/may/10/featuresreviews.guardianreview24

Library of Congress. (n.d.). Fascinating facts. Retrieved February 14, 2026, from https://www.loc.gov/about/fascinating-facts/

Lucas, H. C., Jr., & Goh, J. M. (2009). Disruptive technology: How Kodak missed the digital photography revolution. The Journal of Strategic Information Systems, 18(1), 46-55. https://doi.org/10.1016/j.jsis.2009.01.002

Lupton, M. J. (1998). Maya Angelou: A critical companion. Greenwood Press. https://doi.org/10.5040/9798216977667

Machemer, T. (2021, March 25). This high schooler invented color-changing sutures to detect infection. Smithsonian Magazine. https://www.smithsonianmag.com/innovation/high-schooler-invented-color-changing-sutures-detect-infection-180977345/

Mark, G., Gudith, D., & Klocke, U. (2008). The cost of interrupted work: More speed and stress. In Proceedings of the SIGCHI Conference on Human Factors in Computing Systems (pp. 107–110). Association for Computing Machinery. https://doi.org/10.1145/1357054.1357072

Mazzei, P. (2023, April 20). Florida emergency alert test wakes up residents at 4:45 a.m. The New York Times. https://www.nytimes.com/2023/04/20/us/florida-emergency-alert.html

Meadows, D. H. (2008). Thinking in systems: A primer (D. Wright, Ed.). Chelsea Green Publishing. https://www.chelseagreen.com/product/thinking-in-systems/

Mellow, J. R. (1974). Charmed circle: Gertrude Stein & company. Praeger.

Mellow, J. R. (1992). Hemingway: A life without consequences. Houghton Mifflin.

Merriam-Webster. (n.d.). Prig. In Merriam-Webster.com dictionary. Retrieved February 20, 2026, from https://www.merriam-webster.com/dictionary/prig

Microsoft. (2022). Microsoft new future of work report 2022 (Microsoft Research Technical Report No. MSR-TR-2022-3). https://aka.ms/nfw2022

Microsoft. (2024, July 20). Helping our customers through the CrowdStrike outage. Microsoft Official Blog. https://blogs.microsoft.com/blog/2024/07/20/helping-our-customers-through-the-crowdstrike-outage/

Miller, P. (2013, May 1). I'm still here: Back online after a year without the internet. The Verge. https://www.theverge.com/2013/5/1/4279674/im-still-here-back-online-after-a-year-without-the-internet

Monroe, M. (2010). Fragments: Poems, intimate notes, letters (S. Buchthal & B. Comment, Eds.). Farrar, Straus and Giroux.

Mrazek, M. D., Franklin, M. S., Phillips, D. T., Baird, B., & Schooler, J. W. (2013). Mindfulness training improves working memory capacity and GRE performance while reducing mind wandering. Psychological Science, 24(5), 776–781. https://doi.org/10.1177/0956797612459659

Nash, D. E., Sr. (2018). Kesternich: The battle that saved the Bulge. Army History, (109), 34–50.

National Transportation Safety Board. (2019). Collision between U.S. Navy destroyer John S. McCain and tanker Alnic MC, Singapore Strait, 5 miles northeast of Horsburgh Lighthouse, August 21, 2017 (Marine Accident Report No. NTSB/MAR-19/01). https://www.ntsb.gov/investigations/AccidentReports/Reports/MAR1901.pdf

National Transportation Safety Board. (2019). Collision between US Navy destroyer John S. McCain and tanker Alnic MC, Singapore Strait, 21 August 2017 (Report No. MAR-19/01). https://www.ntsb.gov/investigations/AccidentReports/Reports/MAR1901.pdf

Newport, C. (2016). Deep work: Rules for focused success in a distracted world. Grand Central Publishing.

Newport, C. (2021). A world without email: Reimagining work in an age of communication overload. Portfolio.

Nielsen, J. (2006, April 16). F-shaped pattern for reading web content (Original Study). Nielsen Norman Group. https://www.nngroup.com/articles/f-shaped-pattern-reading-web-content/

Nielsen, J. (2006, April 16). F-shaped pattern for reading web content. Nielsen Norman

Group. https://www.nngroup.com/articles/f-shaped-pattern-reading-web-content/

O'Connor, C. (2012, March 7). Undercover billionaire: Sara Blakely joins the rich list thanks to Spanx. Forbes. https://www.forbes.com/sites/clareoconnor/2012/03/07/undercover-billionaire-sara-blakely-joins-the-rich-list-thanks-to-spanx/

Offerman, N. (2013). Paddle your own canoe: One man's fundamentals for delicious living. Dutton.

Offerman, N. (2023, June 15). Why I love woodworking and hands-on work. The New York Times. https://www.nytimes.com/2023/06/15/arts/offerman-woodworking-parks-recreation.html

Palahniuk, C. (1996). Fight club. W. W. Norton & Company.

Parametrix. (2024). CrowdStrike's impact on the Fortune 500. Parametrix Analytics.

Pasachoff, N. (1996). Marie Curie and the science of radioactivity. Oxford University Press.

Pattakos, A., & Dundon, E. (2017). Prisoners of our thoughts: Viktor Frankl's principles for discovering meaning in life and work (3rd ed.). Berrett-Koehler Publishers.

PBS. (2026). George Eastman | American experience. https://www.pbs.org/wgbh/americanexperience/features/george-eastman/

Perlow, L. A. (2012). Sleeping with your smartphone: How to break the 24/7 habit and change the way you work. Harvard Business Review Press.

Peter, L. J., & Hull, R. (1969). The Peter principle: Why things always go wrong. William Morrow & Company.

Plant, J. (2019). Primitive technology: A survivalist's guide to building tools, shelters, and more in the wild. Clarkson Potter.

Plant, J. (n.d.). Primitive Technology [YouTube channel]. YouTube. Retrieved February 14, 2026, from https://www.youtube.com/channel/UCAL3JXZSzSm8AlZyD3nQdBA

Plimpton, G. (1958). The art of fiction No. 21: Ernest Hemingway. The Paris Review, (18). https://www.theparisreview.org/interviews/4825/the-art-of-fiction-no-21-ernest-hemingway

Profectus Magazine. (2023, October 6). George Eastman: The greatest technology entrepreneur in US history? (An American original). https://profectusmag.com/american-original-george-eastman/

Quinn, S. (1995). Marie Curie: A life. Simon & Schuster.

Remus, D., & Levy, F. (2017). Can robots be lawyers? Computers, lawyers, and the practice of law. The Georgetown Journal of Legal Ethics, 30(3), 501–558.

Rhodes, R. (2011). Hedy's folly: The life and breakthrough inventions of Hedy Lamarr, the most beautiful woman in the world. Doubleday.

Rosa, H. (2013). Social acceleration: A new theory of modernity (J. Trejo-Mathys, Trans.). Columbia University Press. (Original work published 2005) https://doi.org/10.7312/rosa14834

Rosa, H. (2013). Social acceleration: A new theory of modernity (J. Trejo-Mathys, Trans.). Columbia University Press. https://doi.org/10.7312/rosa14834

Saujani, R. (2019). Brave, not perfect: Fear less, fail more, and live bolder. Crown Currency.

Schappell, E., & Brodsky Lacour, C. (1993). The art of fiction No. 134: Toni Morrison. The Paris Review, (128). https://www.theparisreview.org/interviews/1888/the-art-of-fiction-no-134-toni-morrison

Securities and Exchange Commission. (2013). Order instituting administrative and cease-and-desist proceedings: In the matter of Knight Capital Americas LLC (Release No. 70694). https://www.sec.gov/files/litigation/admin/2013/34-70694.pdf

Sellen, A. J., & Harper, R. H. R. (2002). The myth of the paperless office. MIT Press.

Seneca, L. A. (2010). Anger, mercy, revenge (R. A. Kaster & M. C. Nussbaum, Trans.). University of Chicago Press. https://doi.org/10.7208/chicago/9780226748535.001.0001

Sinek, S. (2014). Leaders eat last: Why some teams pull together and others don't. Portfolio.

Sinsky, C., Colligan, L., Li, L., Prgomet, M., Reynolds, S., Goeders, L., Westbrook, J., Tutty, M., & Blike, G. (2016). Allocation of physician time in ambulatory practice: A time and motion study in 4 specialties. Annals of Internal Medicine, 165(11), 753–760. https://doi.org/10.7326/M16-0961

Sonnentag, S. (2012). Psychological detachment from work during leisure time: The benefits of mentally disengaging from work. Current Directions in Psychological Science, 21(2), 114–118. https://doi.org/10.1177/0963721411434979

Sonnentag, S., & Fritz, C. (2007). The Recovery Experience Questionnaire: Development and validation of a measure for assessing recuperation and unwinding from work. Journal of Occupational Health Psychology, 12(3), 204–221. https://doi.org/10.1037/1076-8998.12.3.204

Sonnentag, S., & Fritz, C. (2015). Recovery from job stress: The stressor-detachment model as an integrative framework. Journal of Organizational Behavior, 36(S1), S72–S103. https://doi.org/10.1002/job.1924

Southey, R. (1796). Joan of Arc: An epic poem. Joseph Cottle.

Sparrow, B., Liu, J., & Wegner, D. M. (2011). Google effects on memory: Cognitive consequences of having information at our fingertips. Science, 333(6043), 776–778. https://doi.org/10.1126/science.1207745

Stein, G. (1937). Everybody's autobiography. Random House.

Steiner, R. (1947). Knowledge of the higher worlds and its attainment (G. Metaxa, Trans.). Anthroposophic Press. (Original work published 1904)

Steiner, R. (1994). How to know higher worlds: A modern path of initiation (C. Bamford, Trans.). Anthroposophic Press. (Original work published 1904)

Stosny, S. (2017). Headline stress disorder. Psychology Today. https://www.psychologytoday.com/us/blog/anger-in-the-age-entitlement/201705/headline-stress-disorder

Stosny, S. (2017, March 4). Overcoming headline stress disorder. Psychology Today. https://www.psychologytoday.com/us/blog/anger-in-the-age-entitlement/201703/headline-stress-disorder

strangerkar. (n.d.). Good morning snoopy dance [GIF]. Tenor. https://tenor.com/view/good-morning-snoopy-dance-strangerkar-morning-quotes-gif-3779323093815929635

Susskind, R. (2019). Online courts and the future of justice. Oxford University Press. https://doi.org/10.1093/oso/9780198838364.001.0001

Taylor, F. W. (1911). The principles of scientific management. Harper & Brothers.

Tononi, G., & Cirelli, C. (2014). Sleep and the price of plasticity: From synaptic and cellular homeostasis to memory consolidation and integration. Neuron, 81(1), 12– https://doi.org/10.1016/j.neuron.2013.12.025

Trollope, A. (1883). An autobiography. William Blackwood and Sons. https://www.gutenberg.org/ebooks/5978

Tsao, J. Y., Saunders, H. D., Creighton, J. R., Coltrin, M. E., & Simmons, J. A. (2010). Solid-state lighting: An energy-economics perspective. Journal of Physics. D, Applied Physics, 43(35), 354001. Advance online publication. https://doi.org/10.1088/0022-3727/43/35/354001

Tufekci, Z. (2022, December 31). The Southwest meltdown was a failure of imagination. The New York Times. https://www.nytimes.com/2022/12/31/opinion/southwest-airlines-computers.html

U.S. Department of Transportation. (2023, December 18). DOT penalizes Southwest Airlines $140 million for 2022 holiday meltdown [Press release]. https://www.transportation.gov/briefing-room/dot-penalizes-southwest-airlines-140-million-2022-holiday-meltdown

Vance, A. (2023, January 25). How to be 18 years old again for only $2 million a year. Bloomberg Businessweek. https://www.bloomberg.com/news/features/2023-01-25/anti-aging-techniques-taken-to-extreme-by-bryan-johnson

Von Drehle, D. (2003). Triangle: The fire that changed America. Atlantic Monthly Press.

Wagner-Martin, L. (1995). Favored strangers: Gertrude Stein and her family. Rutgers University Press.

Weil, S. (2009). Waiting for God: Classic essays on spirituality and faith from the philosopher Simone Weil (E.

Craufurd, Trans.). Harper Perennial Modern Classics. (Original work published 1950)

Wolf, M. (2018). Reader, come home: The reading brain in a digital world. HarperCollins.

Woollett, K., & Maguire, E. A. (2011). Acquiring the Knowledge of London's layout drives structural brain changes. Current Biology : CB, 21(24), 2109–2114. https://doi.org/10.1016/j.cub.2011.11.018

Wu, T. (2016). The attention merchants: The epic scramble to get inside our heads. Alfred A. Knopf.

Yates, F. A. (1966). The art of memory. Routledge & Kegan Paul.

Zaleski, O., Huet, E., & Stone, B. (2017, September 8). Inside Juicero's demise, from prized startup to fire sale. Bloomberg. https://www.bloomberg.com/news/features/2017-09-08/inside-juicero-s-demise-from-prized-startup-to-fire-sale

Zeigarnik, B. (1927). Das Behalten erledigter und unerledigter Handlungen [The retention of completed and uncompleted actions]. Psychologische Forschung, 9, 1–85

INDEX

E

F

G

H

T

Marginalia Notes

www.ingramcontent.com/pod-product-compliance
Lightning Source LLC
LaVergne TN
LVHW090357210926
848480LV00002B/367